# BRITISH RAIL STANDARD TANKS

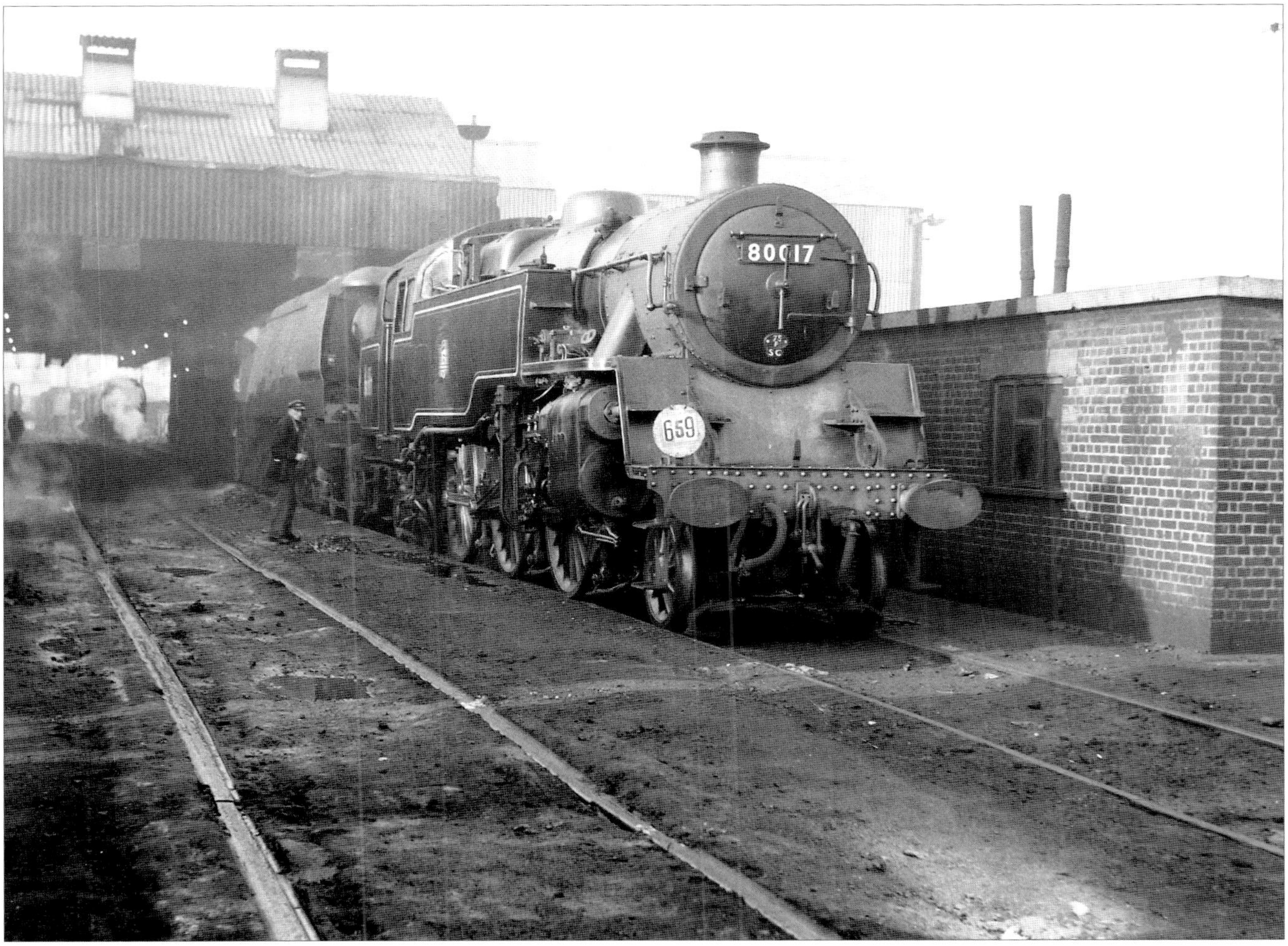

Images from the Transport Treasury archive

## Compiled by Alan C. Butcher

The
· Transport ·
Treasury

I

# The Transport Treasury

## *Reviving the memories of yesterday…*

© Images and design: The Transport Treasury 2025, Text Alan C. Butcher
ISBN 978-1-913251-99-4
First published in 2025 by Transport Treasury Publishing Ltd., 16 Highworth Close, High Wycombe, HP13 7PJ
www.ttpublishing.co.uk
Printed by Short Run Press, 25 Bittern Rd, Exeter EX2 7LW

**The copyright holders hereby give notice that all rights to this work are reserved.**
**Aside from brief passages for the purpose of review, no part of this work may be reproduced, copied by electronic or other means, or otherwise stored in any information storage and retrieval system without written permission from the Publisher. This includes the illustrations herein which shall remain the copyright of the copyright holder.**

Front cover. On 30th June 1962, what the photographer recorded as the Pwllheli portion of 'Cambrian Coast Express', is seen at Dovey Junction behind Class 3MT No 82009. The locomotive entered traffic on 13th June 1952 following completion at Swindon Works that saw it allocated to Tyseley. A nomadic existence saw its arrival at Machynlleth depot early in 1961. A short period allocated to Shrewsbury, in the summer of 1959, may have seen it in action on the Cambrian route earlier in its existence. After a final move to Patricroft, in late March/early April 1965 it was withdrawn on 5th November 1966. *W. A. C. Smith (WS6209)*

Rear cover. Class 4MT No 80059 is seen leaving Elstree tunnel with the 10.55am Bedford-London St Pancras service on 28th May 1953. The locomotive had arrived at Kentish Town depot on 3rd March 1953 following its construction at Brighton Works. In September 1956 it was moved to Chester depot. It returned to London, via Bangor, when it was loaned to Neasden depot in October 1958. December 1959 saw it reallocated to the Southern Region's Dover Marine depot. After a period at Ashford and Tonbridge it arrived in the West Country in June 1962, being allocated to Exmouth Junction. Further moves via Templecombe and Bristol Barrow Road saw its arrival at Bath Green Park from where it was withdrawn on 18th November 1965. It arrived at Buttigiegs (Newport) scrapyard for disposal during January 1966. *A. R. Carpenter*

Title page. Standard Class 4MT No 80017 stands at Bricklayers Arms depot on 25th November 1954. The locomotive was ex-Brighton Works on 8th October 1951 and was allocated to Tunbridge Wells West. It left for Brighton early in 1952, then returned via Eastbourne in late summer the same year. It was withdrawn from Eastleigh depot on 13th September 1964 – a little short of 13 years in traffic. The number on the route indicator disc was the Duty Number – 659 being the 8.26am passenger service from Tunbridge Wells West. Running via East Grinstead it would arrive at London Bridge station at 10.06am. Following a break at 'The Brick' it would return from the capital with the 3.55pm London Bridge-Tunbridge Wells West service. *E. Sawford (ES1970)*

Right. Standard Class 3MT No 82041 brings the cross-country 9.03am Bristol-Bournemouth service into Bath Green Park on 21st July 1959. A product of Swindon Works, No 82041 entered traffic at Barry shed on 17th June 1955. It was reallocated to Bristol Bath Road in early summer 1958. A final transfer to Bath Green Park took place in March 1959 from where it was withdrawn on 31st December 1965. The leading carriage, No 13118E, is an ex-LNER Gresley Vestibule Third Corridor; dating from 1936, it was withdrawn in February 1962. Third class rail travel was abolished from 3rd June 1956 when it was reclassified as Second. *S. Summerson (SUM458)*

Additional thanks go to the following people for their help with the accuracy of this book:
Andrew Penkville          Peter Halsall
Alasdair Taylor           William Armstrong
Malcolm Wells             Jack Kernahan
Bob Allen                 Max Birchenough
Stuart Johnson            Dennis Troughton

# Contents

# Introduction

### The Class 4MT 2-6-4Ts

The initial proposals for 'heavyweight' Class 4 tank locomotives envisaged the adoption of the London, Midland & Scottish (LMS) Railway's Fairburn 2-6-4T design; this was a development of a Stanier design. Like the Ivatt Class 4 and Class 2 Moguls, and the latter's tank derivatives, the Fairburn design incorporated all of the features for easier maintenance that Riddles wanted. At first it appeared that the adoption of BR Standard fittings and 'cosmetic' changes to the cab and front running board would suffice. However, it became clear when design work proceeded at Brighton Locomotive Works that a more drastic approach would be desired in order to fit the BR universal loading gauge – particularly to the cylinders and superstructure. The results were smaller 18in by 28in cylinders, together with slightly smaller driving wheels (5ft 8in) and a boiler at a higher (225lb/sq in) pressure. The redesigned cab and curved side tanks – to take maximum advantage of the loading gauge – gave the locomotives a distinctive front-end appearance.

The majority of the class were constructed at Brighton, the exceptions were – Nos 80000-10 and 80054-8 built at Derby in 1953/4, along with Nos 80106-115 emerging from Doncaster in 1954. The Brighton-built locomotives were constructed in 12 different batches between 1951 and 1957. In all 155 examples were constructed.

As the 1950s turned into the 1960s, locomotives were displaced from their original suburban passenger services by modernisation,

and ended up on less suitable duties. Those locomotives allocated to the Southern Region remained on suburban and cross-country work longer than almost anywhere else, despite successive electrification schemes.

The first of the class to be withdrawn, No 80103, was consigned to the breakers at Stratford Works following its withdrawal from Tilbury depot on 29th August 1962 – a life of under seven and a half years. The Class 4s proved too good to waste and when displaced put in appearances, in the mid-1960s, on long distance cross-country routes including the Central Wales and Somerset & Dorset lines.

Only one locomotive initially survived to be preserved, No 80002 on the Keighley & Worth Valley Railway. Courtesy of Woodham's scrapyard in Barry, South Wales, another 14 examples are still in existence.

### Class 3MT 2-6-2Ts

The need for a Class 3 2-6-2T design was dictated by the existence at Nationalisation of several routes with a maximum axle load of 16 tons – effectively barring them to the 'heavyweight' Class 4 tanks. There was no suitable modern pre-Nationalisation design so they were designed as a scaled-down Class 4, employing basically the same frame design, 17.5in by 26in cylinders with 5ft 3in driving wheels. A lighter boiler, pressed at 200lb/sq in, was derived from the Great Western's Swindon No 2 design used on its '5100', '5600'

**No 80020 is seen on shed carrying a 66A (Polmadie) shed plate, yet according to various records the locomotive was never allocated there! A product of Brighton Works, No 80020 entered traffic on 27th October 1951, making the long journey to the Scottish Region's Kittybrewster depot. It moved to Ardrossan in June 1961, staying for less than a month before arriving at Corkerhill (67A) depot in July. Maybe a fitter got his numbers crossed when replacing the shed plate. The locomotive survived in traffic until withdrawal on 26th June 1965 – it was sold to Motherwell Machinery & Scrap who dismantled the locomotive at its Wishaw depot that September.**

# BR Standard Class 4MT 2-6-4Ts

| No | To Traffic | Date Withdrawn | First shed | Final Shed | No | To Traffic | Date Withdrawn | First shed | Final Shed |
|---|---|---|---|---|---|---|---|---|---|
| 80000 | 29/09/52 | 31/12/66 | Ayr | Corkerhill | 80078 | 02/02/54 | 24/07/65 | Plaistow | Croes Newydd |
| 80001 | 14/10/52 | 17/07/66 | Polmadie | Polmadie | 80079 | 05/03/54 | 24/07/65 | Plaistow | Croes Newydd |
| 80002 | 17/10/52 | 01/03/67 | Motherwell | Polmadie | 80080 | 18/03/54 | 24/07/65 | Plaistow | Croes Newydd |
| 80003 | 24/10/52 | 06/03/65 | Motherwell | St Margarets | 80081 | 13/03/54 | 08/06/65 | Bletchley | Bournemouth |
| 80004 | 04/11/52 | 01/05/67 | Kittybrewster | Corkerhill | 80082 | 15/04/54 | 04/09/66 | Bletchley | Eastleigh |
| 80005 | 31/11/52 | 09/08/66 | Kittybrewster | Polmadie | 80083 | 03/05/54 | 07/08/66 | Bletchley | Eastleigh |
| 80006 | 24/11/52 | 02/09/66 | Polmadie | St Margarets | 80084 | 15/05/54 | 13/06/65 | Bletchley | Redhill |
| 80007 | 08/12/52 | 17/07/66 | Polmadie | Polmadie | 80085 | 28/05/54 | 09/07/67 | Bletchley | Nine Elms |
| 80008 | 15/12/52 | 13/07/64 | Corkerhill | Corkerhill | 80086 | 14/06/54 | 01/05/67 | Bury | Polmadie |
| 80009 | 30/12/52 | 24/09/64 | Corkerhill | Corkerhill | 80087 | 29/06/54 | 14/06/64 | Bury | Eastleigh |
| 80010 | 10/07/51 | 09/06/64 | Tunbridge Wells | Brighton | 80088 | 14/07/54 | 13/06/65 | Bury | Redhill |
| 80011 | 21/07/51 | 09/07/67 | Tunbridge Wells | Bournemouth | 80089 | 03/08/54 | 02/10/66 | Bury | Nine Elms |
| 80012 | 17/08/51 | 19/03/67 | Tunbridge Wells | Nine Elms | 80090 | 12/08/54 | 27/03/65 | Bury | Dundee |
| 80013 | 28/08/51 | 19/09/66 | Tunbridge Wells | Bournemouth | 80091 | 10/09/54 | 29/11/66 | Kentish Town | Beattock |
| 80014 | 05/09/51 | 02/05/65 | Tunbridge Wells | Eastleigh | 80092 | 24/09/54 | 26/09/66 | Kentish Town | Perth |
| 80015 | 14/09/51 | 09/07/67 | Tunbridge Wells | Nine Elms | 80093 | 11/10/54 | 26/09/66 | Bedford | Perth |
| 80016 | 20/09/51 | 09/07/67 | Tunbridge Wells | Eastleigh | 80094 | 25/10/54 | 31/07/66 | Kentish Town | Feltham |
| 80017 | 08/10/51 | 13/09/64 | Tunbridge Wells | Eastleigh | 80095 | 13/11/54 | 02/10/66 | St Albans | Nine Elms |
| 80018 | 13/10/51 | 11/04/65 | Tunbridge Wells | Eastleigh | 80096 | 23/11/54 | 26/12/65 | Plaistow | Bournemouth |
| 80019 | 20/10/51 | 19/03/67 | Tunbridge Wells | Bournemouth | 80097 | 09/12/54 | 24/07/65 | Plaistow | Machynlleth |
| 80020 | 27/10/51 | 26/06/65 | Kittybrewster | Corkerhill | 80098 | 22/12/54 | 24/07/65 | Plaistow | Machynlleth |
| 80021 | 10/11/51 | 13/07/64 | Kittybrewster | Corkerhill | 80099 | 17/01/55 | 08/05/65 | Plaistow | Machynlleth |
| 80022 | 20/11/51 | 26/06/65 | Polmadie | St Margarets | 80100 | 31/01/55 | 24/07/65 | Plaistow | Shrewsbury |
| 80023 | 27/11/51 | 02/10/65 | Polmadie | Dumfries | 80101 | 07/02/55 | 17/07/65 | Plaistow | Machynlleth |
| 80024 | 07/12/51 | 24/08/66 | Corkerhill | Corkerhill | 80102 | 01/03/55 | 05/12/65 | Plaistow | Eastleigh |
| 80025 | 13/12/51 | 24/08/66 | Corkerhill | Corkerhill | 80103 | 16/03/55 | 29/08/62 | Plaistow | Tilbury |
| 80026 | 20/12/51 | 02/09/66 | Polmadie | St Margarets | 80104 | 31/03/55 | 24/07/65 | Plaistow | Machynlleth |
| 80027 | 04/01/52 | 29/11/66 | Polmadie | Polmadie | 80105 | 19/04/55 | 24/07/65 | Plaistow | Machynlleth |
| 80028 | 11/01/52 | 26/09/66 | Kittybrewster | Perth | 80106 | 22/10/54 | 12/10/64 | Kittybrewster | Polmadie |
| 80029 | 22/01/52 | 16/12/65 | Kittybrewster | Hurlford | 80107 | 29/10/54 | 14/09/64 | Kittybrewster | Polmadie |
| 80030 | 05/02/52 | 08/06/64 | Ayr | Corkerhill | 80108 | 05/11/54 | 08/05/65 | Kittybrewster | Polmadie |
| 80031 | 02/02/52 | 22/09/64 | Brighton | Redhill | 80109 | 18/11/54 | 13/11/65 | Kittybrewster | Polmadie |
| 80032 | 04/03/52 | 29/01/67 | Brighton | Bournemouth | 80110 | 18/11/54 | 08/05/65 | Kittybrewster | Polmadie |
| 80033 | 18/03/52 | 02/10/66 | Brighton | Feltham (stored) | 80111 | 26/11/54 | 29/11/66 | Polmadie | Beattock |
| 80034 | 01/04/52 | 02/01/66 | Crewe North | Feltham (stored) | 80112 | 03/12/54 | 24/08/66 | Polmadie | Corkerhill |
| 80035 | 15/05/52 | 21/04/65 | Watford | Yeovil | 80113 | 14/12/54 | 02/09/66 | Polmadie | St Margarets |
| 80036 | 23/05/52 | 06/11/64 | Watford | Exmouth Junction | 80114 | 29/12/54 | 31/12/66 | Polmadie | St Margarets |
| 80037 | 24/05/52 | 07/03/66 | Watford | Templecombe | 80115 | 31/12/54 | 12/10/64 | Polmadie | Polmadie |
| 80038 | 07/06/52 | 07/09/64 | Watford | Exmouth Junction | 80116 | 04/05/55 | 01/05/67 | York | Polmadie |
| 80039 | 18/06/52 | 21/01/66 | Bletchley | Templecombe | 80117 | 19/05/55 | 03/03/66 | Whitby | Polmadie |
| 80040 | 27/06/52 | 06/05/64 | Bletchley | Exmouth Junction | 80118 | 15/06/55 | 29/11/66 | Whitby | Polmadie |
| 80041 | 05/07/52 | 07/03/66 | Bletchley | Templecombe | 80119 | 21/06/55 | 29/05/65 | Whitby | Dumfries |
| 80042 | 15/07/52 | 06/02/65 | Bletchley | Exmouth Junction | 80120 | 05/07/55 | 01/05/67 | Whitby | Polmadie |
| 80043 | 24/07/52 | 07/03/66 | Bletchley | Templecombe | 80121 | 22/07/55 | 02/06/66 | Kittybrewster | Polmadie |
| 80044 | 21/08/52 | 14/11/64 | Derby | Bangor | 80122 | 08/08/55 | 31/12/66 | Kittybrewster | Greenock Ladyburn |
| 80045 | 04/09/52 | 01/05/67 | Bedford | Polmadie | 80123 | 09/09/55 | 17/08/66 | Dundee Tay Bridge | Polmadie |
| 80046 | 17/09/52 | 01/05/67 | Bedford | Corkerhill | 80124 | 22/09/55 | 10/12/66 | Dundee Tay Bridge | St Margarets |
| 80047 | 30/09/52 | 24/08/66 | Bedford | Corkerhill | 80125 | 06/10/55 | 10/10/64 | Stirling | Lostock Hall |
| 80048 | 13/10/52 | 17/07/65 | Kentish Town | Shrewsbury | 80126 | 20/10/55 | 19/11/66 | Perth | Perth |
| 80049 | 24/10/52 | 08/06/64 | Blackpool | Corkerhill | 80127 | 07/11/55 | 30/07/64 | Corkerhill | Corkerhill |
| 80050 | 06/11/52 | 21/11/64 | Newton Heath | Bangor | 80128 | 23/11/55 | 04/04/67 | Corkerhill | Corkerhill |
| 80051 | 21/11/52 | 17/08/66 | Newton Heath | Corkerhill | 80129 | 09/12/55 | 10/10/64 | Polmadie | Lostock Hall |
| 80052 | 05/12/52 | 07/07/64 | Newton Heath | Corkerhill | 80130 | 22/12/55 | 17/08/66 | Polmadie | Polmadie |
| 80053 | 22/12/52 | 07/07/64 | Newton Heath | Corkerhill | 80131 | 02/03/56 | 08/05/65 | Plaistow | Bangor |
| 80054 | 04/12/54 | 29/06/66 | Polmadie | Greenock Ladyburn | 80132 | 15/03/56 | 09/01/66 | Plaistow | Eastleigh |
| 80055 | 23/12/54 | 02/09/66 | Polmadie | St Margarets | 80133 | 28/03/56 | 09/07/67 | Plaistow | Nine Elms |
| 80056 | 30/12/54 | 10/10/64 | Polmadie | Lostock Hall | 80134 | 20/04/56 | 09/07/67 | Plaistow | Bournemouth |
| 80057 | 31/12/54 | 31/12/66 | Polmadie | Polmadie | 80135 | 30/04/56 | 24/07/65 | Plaistow | Shrewsbury |
| 80058 | 08/01/55 | 17/07/66 | Polmadie | Polmadie | 80136 | 11/05/56 | 24/07/65 | Plaistow | Shrewsbury |
| 80059 | 13/03/53 | 18/11/65 | Kentish Town | Bath Green Park | 80137 | 28/05/56 | 31/10/65* | Neasden | Nine Elms |
| 80060 | 19/03/53 | 07/02/66 | Bedford | Polmadie | 80138 | 12/06/56 | 02/10/66 | Neasden | Bournemouth |
| 80061 | 08/04/53 | 31/12/66 | Bedford | Polmadie | 80139 | 26/06/56 | 09/07/67 | Neasden | Eastleigh |
| 80062 | 29/04/53 | 12/10/64 | Kentish Town | Stirling | 80140 | 10/07/56 | 09/07/67 | Neasden | Nine Elms |
| 80063 | 19/05/53 | 24/08/66 | Saltley | Corkerhill | 80141 | 25/07/56 | 09/01/66 | Neasden | Nine Elms |
| 80064 | 09/06/53 | 25/08/65 | Watford | Bristol Barrow Road | 80142 | 09/08/56 | 03/03/66 | Neasden | Eastleigh |
| 80065 | 25/06/53 | 04/09/66 | Watford | Eastleigh | 80143 | 10/09/56 | 09/07/67 | Neasden | Nine Elms |
| 80066 | 15/07/53 | 13/06/65 | Watford | Eastleigh | 80144 | 25/09/56 | 15/05/66 | Neasden | Nine Elms |
| 80067 | 04/08/53 | 03/06/65 | Watford | Bristol Barrow Road | 80145 | 16/10/56 | 25/06/67 | Brighton | Nine Elms |
| 80068 | 24/08/53 | 02/10/66 | Watford | Feltham (stored) | 80146 | 30/10/56 | 09/07/67 | Brighton | Bournemouth |
| 80069 | 30/09/53 | 23/01/66 | Plaistow | Nine Elms | 80147 | 15/11/56 | 13/06/65 | Brighton | Bournemouth |
| 80070 | 09/10/53 | 20/06/65 | Plaistow | Eastleigh | 80148 | 29/11/56 | 14/06/64 | Brighton | Feltham |
| 80071 | 21/10/53 | 30/07/64 | Plaistow | Carstairs | 80149 | 13/12/56 | 07/03/65 | Brighton | Redhill |
| 80072 | 03/11/53 | 24/07/65 | Plaistow | Shrewsbury | 80150 | 28/12/56 | 17/10/65 | Brighton | Eastleigh |
| 80073 | 13/11/53 | 30/07/64 | Plaistow | Carstairs | 80151 | 18/01/57 | 07/05/67 | Brighton | Eastleigh |
| 80074 | 27/11/53 | 30/07/64 | Plaistow | Carstairs | 80152 | 06/02/57 | 09/07/67 | Brighton | Nine Elms |
| 80075 | 10/12/53 | 30/07/64 | Plaistow | Carstairs | 80153 | 25/02/57 | 07/03/65 | Brighton | Redhill |
| 80076 | 23/12/53 | 30/07/64 | Plaistow | Dumfries | 80154 | 26/03/57 | 02/04/67 | Brighton | Nine Elms |
| 80077 | 12/01/54 | 12/10/64 | Plaistow | Corkerhill | | | | | |

* withdrawn week ending (Saturday)

and '8100' classes. Designed and constructed at Swindon the class made its debut in 1952, with 45 locomotives, Nos 82000-44, built in five batches for the Southern and Western regions over a three year period.

Orders were placed in 1954 for an additional 18 locomotives destined for the Western and North Eastern regions – with the decision to proceed with dieselisation, these were cancelled. The Western Region examples could be seen from South Devon to North Wales. In later years they were also to be seen on the Cambrian Coast lines. The Southern Region examples were used in the West Country – on both the East Devon and North Cornwall 'Withered Arm' line services.

Once displaced from their original services, the Southern allocated examples moved east to work empty stock trains into Waterloo station.

No examples survived to be preserved; however a 'new build' is under construction at the Severn Valley Railway.

### Class 2MT 2-6-2Ts

The Class 2MTs were virtually a straight copy of the Ivatt original design dating from 1946. The main changes were Standard cab fittings and a tidier arrangement of the front end. They had 16.5in by 24in cylinders with 5ft driving wheels. Designed at Crewe the first 20, Nos 84000-19, were built at Crewe in 1953. All were originally fitted with push-pull control apparatus and most were put onto branch line duties replacing antiquated motive power.

An additional batch of 10, Nos 84020-29, left Darlington Works for the Southern Region in 1957 – amazingly the order had been placed in 1953 – again replacing older motive power. With modernisation and branch line closures the locomotives simply ran out of work and withdrawals commenced in 1963 with No 84012 being withdrawn from Southport depot on 12th October – being consigned to Crewe works for scrapping the same month. The record for the shortest working life rests with No 84027 that served for less than seven years – No 84029 served for seven years and two days.

No examples survived to be preserved; however a 'rebuild' is under construction at the Bluebell Railway using the major parts from Class 2MT 2-6-0 No 78059.

### A footnote:

As a photographic record this is not the place to give a complete history of the classes, and those wishing additional details are directed to the RCTS books on the classes – *A Detailed History of British Railways Standard Steam Locomotives: Volumes 3 and 5.*

*Note on dates: The BR practice was to use the 'period ending' dates, using the Saturday as the end date. So unless the actual date is quoted then the dates recorded are either week ending, four ending, or in some cases five or six weeks ending. Rather than scatter the text with numerous 'week/period ending', a looser description is given and readers are referred to the data tables and the RCTS books mentioned above.*

**No 82007 is seen outside Swindon Works following completion in 1952, the locomotive is not carrying a shed plate. Allocated new to Tyseley (84E) it entered traffic on 30th May. It moved to South Wales the following year when it was reallocated to Barry depot. As with many of the class it was nomadic as its day-to-day services were dieselised – including stays at Shrewsbury, Leamington, Wrexham Rhosddu (the town's ex-GCR shed), and various Bristol depots. It was withdrawn from Bristol Barrow Road depot on 29th June 1964, to be recycled that December at Cashmore's Newport scrapyard after just over 12 years service.** *N. Stead Collection (NS207841)*

## BR Standard Class 3MT 2-6-2Ts

| No | To Traffic | Date Withdrawn | First shed | Final Shed |
|---|---|---|---|---|
| 82000 | 04/04/52 | 10/12/66 | Tyseley | Patricroft |
| 82001 | 18/04/52 | 31/12/65 | Tyseley | Bath Green Park |
| 82002 | 28/04/52 | 07/02/64 | Tyseley | Exmouth Junction |
| 82003 | 02/25/52 | 12/12/66 | Tyseley | Patricroft |
| 82004 | 14/05/52 | 01/10/65 | Tyseley | Bath Green Park |
| 82005 | 16/05/52 | 19/09/65 | Tyseley | Nine Elms |
| 82006 | 27/05/52 | 18/09/66 | Tyseley | Nine Elms |
| 82007 | 30/05/52 | 25/06/64 | Tyseley | Bristol Barrow Road |
| 82008 | 05/06/52 | 07/02/64 | Tyseley | Taunton |
| 82009 | 13/06/52 | 05/11/66 | Tyseley | Patricroft |
| 82010 | 25/06/52 | 25/04/65 | Exmouth Junction | Nine Elms |
| 82011 | 01/07/52 | 30/08/64 | Exmouth Junction | Nine Elms |
| 82012 | 04/07/52 | 31/05/64* | Exmouth Junction | Nine Elms |
| 82013 | 09/07/52 | 14/06/64 | Exmouth Junction | Nine Elms |
| 82014 | 01/08/52 | 31/05/64* | Exmouth Junction | Nine Elms |
| 82015 | 13/08/52 | 06/12/64 | Exmouth Junction | Nine Elms |
| 82016 | 21/08/52 | 25/04/65 | Exmouth Junction | Nine Elms |
| 82017 | 08/08/52 | 25/04/65 | Exmouth Junction | Nine Elms |
| 82018 | 08/09/52 | 10/07/66 | Exmouth Junction | Nine Elms |
| 82019 | 25/09/52 | 09/07/67 | Exmouth Junction | Nine Elms |
| 82020 | 29/09/54 | 19/09/65 | Hull Botanic Gardens | Nine Elms |
| 82021 | 06/10/54 | 17/10/65 | Hull Botanic Gardens | Nine Elms |
| 82022 | 15/10/54 | 17/10/65 | Exmouth Junction | Nine Elms |
| 82023 | 22/10/54 | 02/10/66 | Exmouth Junction | Nine Elms |
| 82024 | 29/10/54 | 30/01/66 | Exmouth Junction | Nine Elms |
| 82025 | 05/11/54 | 09/08/64 | Exmouth Junction | Nine Elms |
| 82026 | 12/11/54 | 26/06/66 | Kirkby Stephen | Nine Elms |
| 82027 | 23/11/54 | 09/01/66 | Kirkby Stephen | Nine Elms |
| 82028 | 02/12/54 | 04/09/66 | Darlington | Nine Elms |
| 82029 | 13/12/54 | 09/07/67 | Darlington | Nine Elms |
| 82030 | 21/12/54 | 31/12/65 | Barry | Bath Green Park |
| 82031 | 30/12/54 | 10/12/66 | Barry | Patricroft |
| 82032 | 01/01/55 | 01/05/65 | Barry | Bangor |
| 82033 | 13/01/55 | 19/09/65 | Newton Abbot | Nine Elms |
| 82034 | 28/01/55 | 10/12/66 | Newton Abbot | Patricroft |
| 82035 | 03/03/55 | 06/08/65 | Barry | Yeovil Town |
| 82036 | 01/04/55 | 21/07/65 | Barry | Bristol Barrow Road |
| 82037 | 20/04/55 | 25/08/65 | Swansea Victoria | Bristol Barrow Road |
| 82038 | 04/05/55 | 06/08/65 | Newton Abbot | Bristol Barrow Road |
| 82039 | 10/05/55 | 02/07/65 | Barry | Gloucester Horton Rd |
| 82040 | 19/05/55 | 02/07/65 | Barry | Gloucester Horton Rd |
| 82041 | 17/06/55 | 31/12/65 | Barry | Bath Green Park |
| 82042 | 20/06/55 | 06/08/65 | Barry | Gloucester Horton Rd |
| 82043 | 27/06/55 | 07/02/64 | Barry | Bristol Barrow Road |
| 82044 | 12/08/55 | 18/11/65 | Barry | Bath Green Park |

\* withdrawn week ending (Saturday)

## BR Standard Class 2MT 2-6-2Ts

| No | To Traffic | Date Withdrawn | First shed | Final Shed |
|---|---|---|---|---|
| 84000 | 04/07/53 | 30/10/65 | Crewe North | Croes Newydd |
| 84001 | 06/07/53 | 31/10/64* | Crewe North | Llandudno Junction |
| 84002 | 07/08/53 | 17/04/65 | Plodder Lane | Bletchley |
| 84003 | 10/08/53 | 02/10/65 | Plodder Lane | Llandudno Junction |
| 84004 | 19/08/53 | 30/10/65 | Plodder Lane | Croes Newydd |
| 84005 | 24/08/53 | 30/10/65 | Bedford | Leicester Midland |
| 84006 | 26/08/53 | 30/10/65 | Eurton | Leicester Midland |
| 84007 | 31/08/53 | 18/01/64 | Burton | Wellingborough |
| 84008 | 31/08/53 | 30/10/65 | Burton | Leicester Midland |
| 84009 | 03/09/53 | 20/11/65* | Royston | Llandudno Junction |
| 84010 | 08/09/53 | 11/12/65 | Low Moor | Fleetwood† |
| 84011 | 18/09/53 | 24/04/65 | Low Moor | Fleetwood |
| 84012 | 17/09/53 | 12/10/63 | Low Moor | Southport |
| 84013 | 21/09/53 | 11/12/65 | Low Moor | Stockport Edgeley† |
| 84014 | 22/09/53 | 11/12/65 | Low Moor | Stockport Edgeley† |
| 84015 | 02/10/53 | 11/12/65 | Low Moor | Skipton† |
| 84016 | 07/10/53 | 11/12/65 | Bury | Fleetwood† |
| 84017 | 13/10/53 | 11/12/65 | Bury | Stockport Edgeley† |
| 84018 | 23/10/53 | 24/04/65 | Bury | Fleetwood |
| 84019 | 31/10/53 | 11/12/65 | Bury | Bolton† |
| 84020 | 27/03/57 | 31/10/64 | Ashford* | Llandudno Junction |
| 84021 | 30/03/57 | 05/09/64 | Ashford | Crewe Works |
| 84022 | 30/03/57 | 05/09/64 | Ashford | Crewe Works |
| 84023 | 05/04/57 | 05/09/64 | Ashford | Crewe Works |
| 84024 | 10/04/57 | 05/09/64 | Ashford | Crewe Works |
| 84025 | 16/04/57 | 11/12/65 | Ramsgate | Bolton† |
| 84026 | 18/04/57 | 11/12/65 | Ramsgate | Stockport Edgeley† |
| 84027 | 19/05/57 | 02/05/64 | Ramsgate | Annesley |
| 84028 | 15/05/57 | 11/12/65 | Ramsgate | Skipton† |
| 84029 | 11/06/57 | 13/06/64 | Ramsgate | Leicester Midland |

\* withdrawn week ending (Saturday)

† Transferred to Eastleigh w/e 27/11/65 (for Isle of Wight), returned w/e 04/12/65 to originating depot for immediate withdrawal.
Only No 84014 reached Eastleigh and was despatched to Cashmore's Newport scrapyard from there; the remainder being 'paper' transfers though some may have already have started the trip south

No 84029 is seen outside Darlington Works following its completion in June 1957. It entered traffic on 11th June, being allocated to the Southern Region's Ramsgate depot. After stays at Ashford and Eastleigh it was transferred to the London Midland Region when it arrived at Bedford in September 1961. It arrived at Leicester Midland, its final depot, in February 1963 and was withdrawn on 13th June 1964. The locomotive was in traffic for two days over seven years, meeting its fate at the hands of Cashmore's (Great Bridge) cutters during October 1964.

# The Class of 1951

Top left. No 80011 had been completed at Brighton Works, entering traffic on 21st July 1951 and allocated to Tunbridge Wells West. It is seen at Sanderstead during August 1952 on duty number 656. This turn involved leaving the depot at 6.12am running light engine to Groombridge to operate the 6.18 service to London Bridge. After a days work it would return to its home shed at 7.56pm. It left Tunbridge in November 1956 when it was allocated to Brighton. It returned to Tunbridge, in January 1963, via Three Bridges for a second stint there. Returning to Brighton that September it spent the next four years around various Southern depots before ending its days at Bournemouth depot. Surviving until the last day of steam on the Southern Region, 9th July 1967, it was sold to Birds, Risca, for demolition. The goods yard here is already looking unkempt, although it was not to officially close until 20th March 1961. *J. Flint/J. Harbart (FH970)*

Bottom left. The second of the first batch of locomotives from Brighton was sent to Tunbridge Wells West when it entered traffic on 17th August 1951, and is seen here on shed a few weeks later. The original two-road locomotive shed, dating from 1st October 1866, was capable of accommodating six engines and situated to the south of the station. A larger four-road shed located to the north west of the station replaced it in 1890. Following bomb damage during World War 2, on 20th November 1940, the slate roof of the engine shed was replaced with corrugated asbestos. It was closed to steam locomotives on 9th September 1963 but the tracks were used for storage of stock until the late 1970s when they were removed. Today it serves the locomotive department of the Spa Valley Railway. No 80012 was not so lucky and was withdrawn from Nine Elms depot on 19th March 1967. *J. Flint/J. Harbart (FH960)*

Above. For those who are not 'Men of Kent' confusion can arise between Tonbridge and Tunbridge – particularly as they are close together! No 80016 is seen here on 3rd September 1955 leaving Tonbridge for Tunbridge whilst allocated to Brighton depot. New to Tunbridge Wells West on 20th September 1951, it was reallocated to Brighton in March 1952, only to return in November 1956 – staying until the depot closed to steam when it returned to Brighton. A final move to Eastleigh occurred in June 1964 when it was one of the Standard Fours to survive until the end of steam on the Southern Region, being withdrawn on 9th July 1967. *J. Flint/J. Harbart (FH908)*

# British Railways Standard Tanks

Below. No 80018 stands against the buffer stops at Victoria having brought a local service in, what appears to be a Fairburn Class 4MT 2-6-4T has already attached itself to the rear of the train to take the next service out. No 80018 entered service on 13th October 1951, being allocated to Tunbridge Wells West where it remained three months until transfer to Brighton. With stints at Eastbourne and Redhill as well as reallocations to Brighton and Tunbridge it eventually left the area in the summer of 1964 when it was reallocated to Feltham. A final move to Eastleigh in November the same year saw its withdrawal on 11th April 1965; four months later Cohens at Morriston cut it up.

Top right. No 80019 is seen running into Lewes station from the east, passing an impressive signal gantry as it does so, at the head of a cross-country service. The locomotive entered service on 20th October 1951, although not officially allocated to Brighton depot, adjacent to its construction site, until January 1952. Brighton depot was formally closed to steam on 15th June 1964, but remained in use for stabling steam locomotives until 1964 and was demolished on the 14th August 1966. No 80019 left the Central Division of the Southern Region in June 1965 when it was reallocated to Bournemouth from where it was withdrawn on 19th March 1967 and dispatched to Buttigiegs, Newport, for recycling.

Bottom right. No 80021 entered traffic on 10th November 1951 and made the long journey from Brighton Works to Kittybrewster. It is seen soon after arrival as it has yet to have its 61A shed plate fitted. It stayed at Kittybrewster for almost 10 years, being reallocated to Ardrossan in June 1961 with a final move to Corkerhill that July. It was withdrawn three years later on 13th July 1964. In the background is a Keith allocated Class B12, a Stratford (London) product of Great Eastern Railway design dating from March 1913. No 61507 would be withdrawn on 26th February 1953 having served three owners for almost 40 years. The original depot at Kittybrewster was opened by the Great North of Scotland Railway on 12th September 1854. It was partially rebuilt by the LNER and later reroofed by the Scottish Region; it closed to steam on 12th June 1961 and was used for diesel traction for a few years before demolition. *H. S. Brighty*

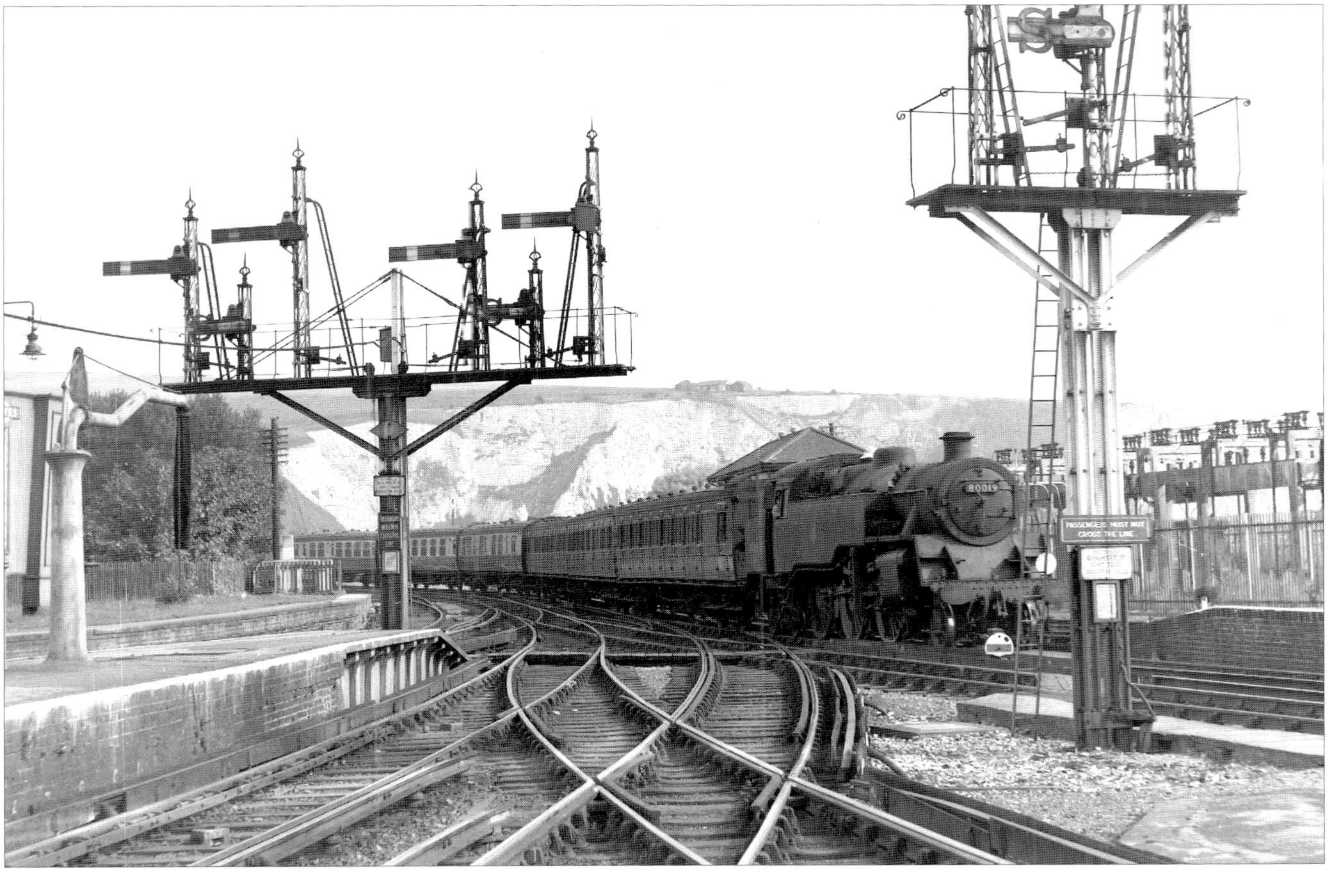
80019
CROSS THE LINE

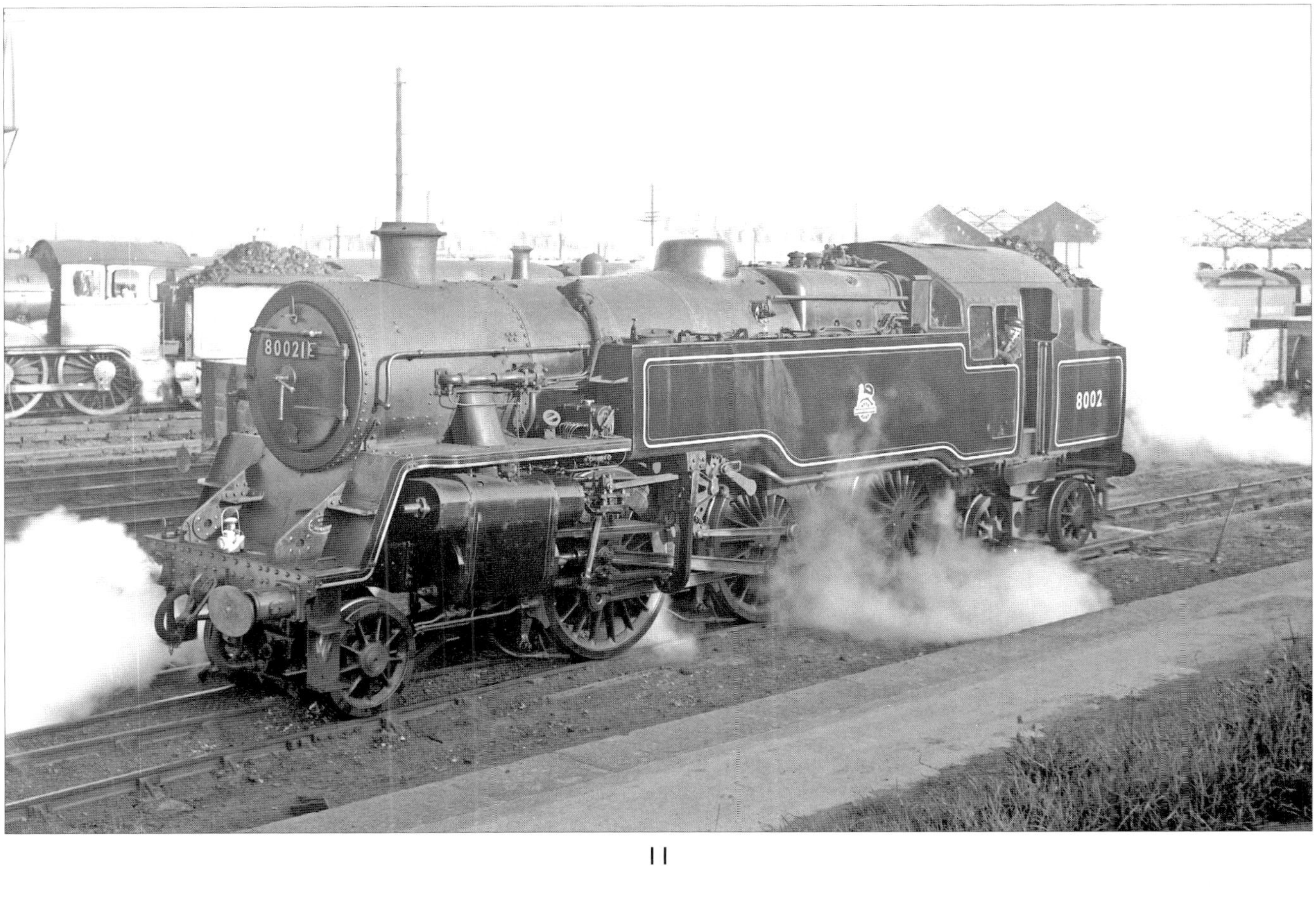
80021E
8002

# The Class of 1952

Top left. No 80024 is seen climbing Neilston Bank to the south-west of Glasgow on the former Glasgow Barrhead & Kilmarnock line which extended to Kilmarnock the line which originally terminated at Crofthead (Neilston). The track over the Bank was singled in 1973. No 80024 was allocated to Corkerhill on 7th December 1951 making the long journey from Brighton – this was its only depot being withdrawn on 24th August 1966. With a six coach rake of suburban stock in tow, the locomotive carries the electrification warning signs. Following withdrawal No 80024 met it fate at the hands of Shipbreaking Industries cutters at Faslane during October the same year – a final resting place for a number of Scottish based Class 4 tanks.

Bottom left. No 80026 departs from Edinburgh Princes Street with a train for Leith on 23rd February 1952, a little over eight weeks after arriving at Polmadie from Brighton – the shine has already gone off the paintwork. Princes Street station stood at the west end of Princes Street, for almost 100 years. Temporary stations were opened in 1848 and 1870, with construction of the main station by the Caledonian Railway commencing in the 1890s. The station served the main line to London, via Carstairs, heading south-west from the station, which was later augmented with a number of suburban stops, Merchiston, Slateford, and Kingsknowe, and a branch line to Colinton and Balerno. The Caledonian Railway later added other suburban lines serving the north and west of the city, including Barnton, Davidson's Mains, Granton, and Leith. The station was closed completely in 1965 and largely demolished in 1969-70. Only its hotel remains, but it is no longer in railway ownership. No 80026 remained at Polmadie until the summer of 1962 when it was reallocated to St Margarets from where it was withdrawn on 2nd September 1966.

Above. Nominally the first of the class, although not the first to enter service, No 80000 entered traffic from Derby Works on 26th September 1952 and was allocated to Ayr, with an almost immediate reallocation to Corkerhill. A move to Hurlford took place in October 1961 with a return to Corkerhill in January 1962 from where it was withdrawn on 31st December 1966. Sold to Shipbreaking Industries, it was cut up on site rather than make the journey to Faslane. It is seen here near Dreghorn on the line between Irvine and Kilmarnock – the line closed to passenger services on 6th April 1964 and totally the following year. *N. Stead collection (207844)*

# The Class of 1952

Top left. No 80001 leaving the Cathcart engine shunting siding for Platform 8 at Glasgow Central on 15th September 1955. This siding was used by engines released from an incoming Cathcart Circle service before moving to the adjacent platform to take out the next Circle service. No 80001 departed Derby Works bound for Scotland on 14th October 1952. It stayed at Polmadie until reallocated to Beattock during early summer 1962. It returned to Polmadie in May 1964 to see out its final two years in service, being withdrawn on 17th July 1966, to be broken up at Faslane alongside numerous classmates that October.

Bottom left. A grimy No 80003 is seen passing through Edinburgh Princes Street Gardens at the head of a four-coach train of BR Mark 1 suburban stock. These are two adjacent public parks in the centre of the city, lying in the shadow of Edinburgh Castle. The gardens were created in the 1820s following the draining of Nor Loch and building of the New Town, beginning in the 1760s. The loch was an artificial creation situated on the north side of the town, the water was habitually polluted and drained downhill to the old town. New to traffic on 24th October 1952 No 80003 was initially allocated to St Margarets depot, departing for Polmadie within days of arrival. After almost 10 years it was sent back to St Margarets from where it was withdrawn on 6th March 1965. *I. Strachan*

Below. Released to traffic from Derby Works on 4th November 1952 No 80004 was allocated to Kittybrewster shed, where it is seen on 11th May 1954. It was reallocated to Beattock in June 1961, moving via Eastfield and Dawsholm to Corkerhill from where it was withdrawn on 1st May 1967. Note the recessed panel in the cab side that was used to accommodate automatic tablet-catching apparatus. The works of the Great North of Scotland Railway was originally at Kittybrewster, opening in 1857, on the west side of the station. In a cramped location, it was relocated to a green field site at Inverurie in 1901. The GNSR lines were amongst the earliest to be fully dieselised and the shed closed to steam on 12th June 1961 with most locomotives being transferred to Ferryhill. Today nothing remains of the works or engine shed. *A. W. Battson*

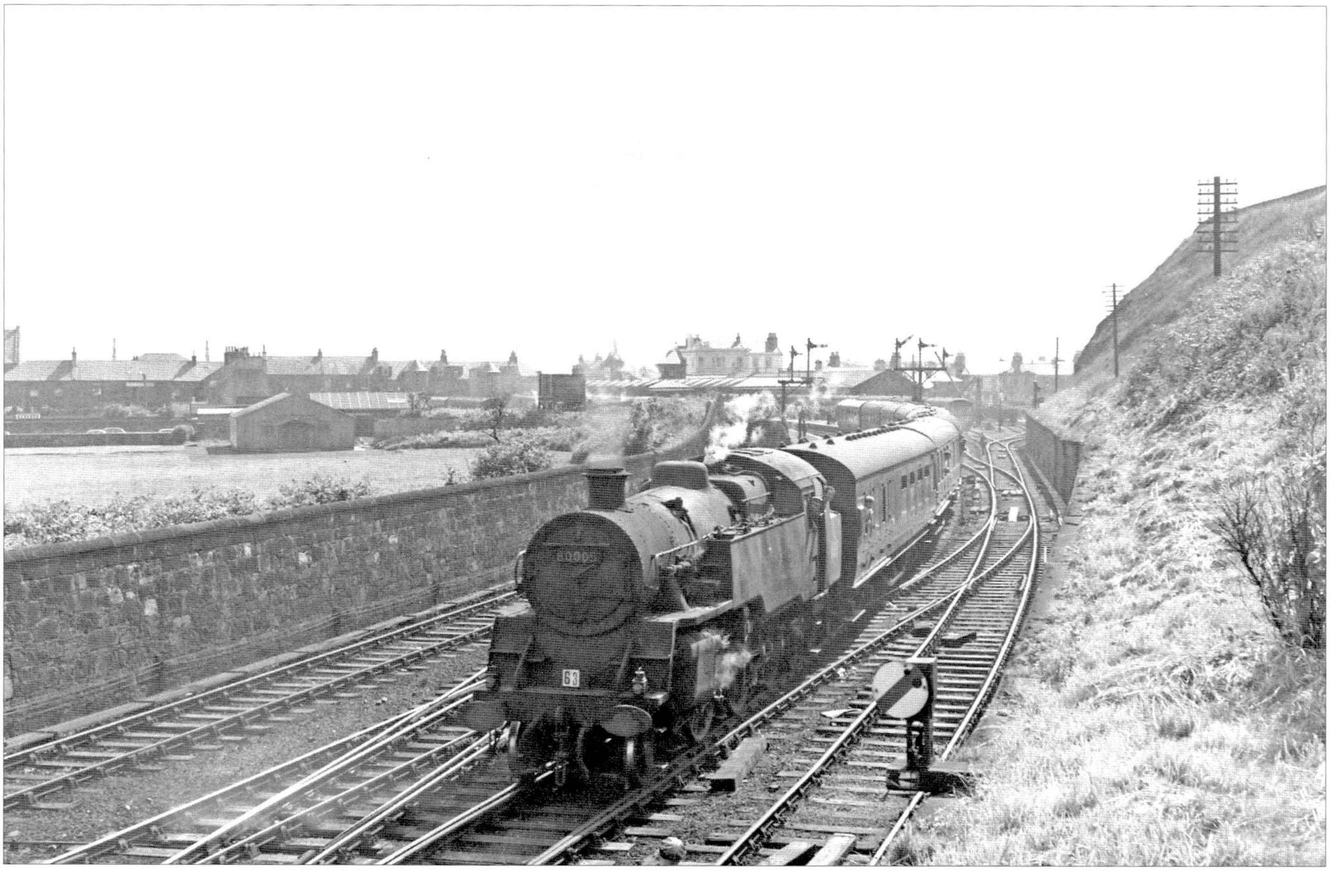

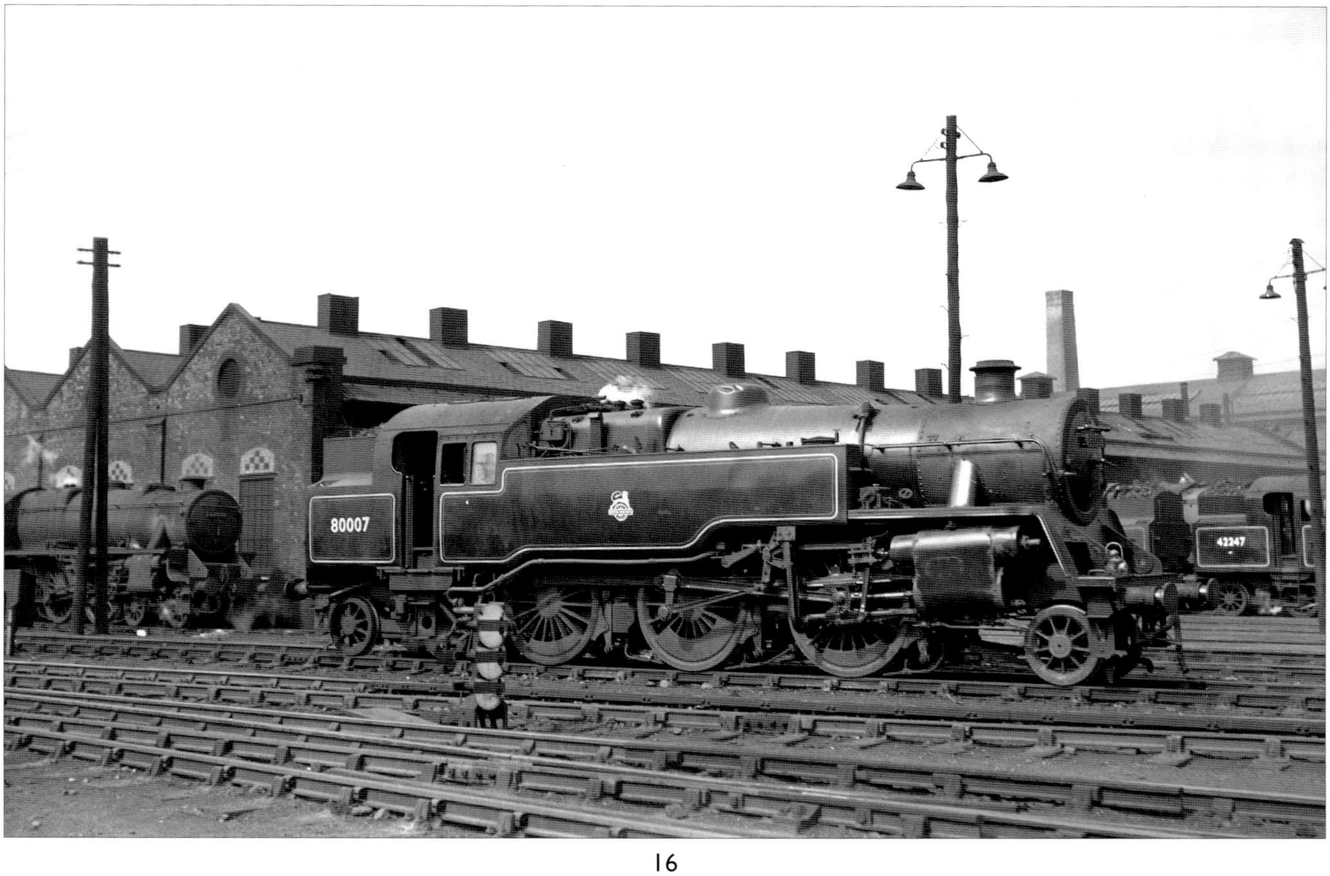
80007
42247

Top left. No 80005 has just left Ardrossan Town station with a Summer Extra on 20th July 1964 whilst allocated to Corkerhill depot. It entered service on 13th November 1952 allocated to Kittybrewster. It was transferred to Corkerhill in July 1959 before moving to Ayr in October 1964. March 1965 saw a reallocation to Beattock where it remained until that November when a final move to Polmadie took place. Withdrawal occurred on 9th August 1966. The station was opened in 1831 by the Ardrossan Railway and was simply known as Ardrossan. The original station had two side platforms and although it was a terminus at first it became an intermediate station upon the opening of Ardrossan Pier railway station in 1840. The station was rebuilt some time around 1890. It became part of the Glasgow & South Western Railway, passing to the LMS at the Grouping of 1923. The station then passed on to the Scottish Region of British Railways on nationalisation in 1948 and was renamed Ardrossan Town by BR on 2nd March 1953. The station was closed on 1st January 1968 and lay derelict for a number of years, though the double tracks into the bay platform remained and were used for DMU storage. Upon electrification of the Ayrshire Coast Line, the station was reopened on 19th January 1987. *W. A. C. Smith (WS7557)*

Bottom left. No 80007 is seen on shed at Polmadie on 8th May 1954 during its first stint at the depot, it had arrived from Derby in early December 1952. The original Polmadie depot was opened by the Polloc & Govan Railway, later taken over by the Caledonian Railway. It was rebuilt several times, notably by the LMS in 1925 when it had 14 tracks in it. Nationalisation saw the shed yard remodelled and the structure reclad in corrugated sheeting. It closed to steam on 1st May 1967 and saw use as a diesel depot for a number of years. No 80007 was transferred to St Margarets during early summer of 1962 before a return to Polmadie in November 1965. It was withdrawn on 17th July 1966, meeting it fate at Faslane in October the same year.

Above. A smart looking No 80009 stands under the coaling plant at Corkerhill depot on 5th June 1954. It had entered traffic on 30th December 1952 and was allocated here where it was to remain for its entire working life. Withdrawal came on 24th September 1964, and sold to Motherwell Machinery & Scrap, Wishaw, for recycling. The Glasgow & South Western Railway opened the six-road depot in 1896. Its facilities included a ramped coal stage with water tank over, along with a turntable and repair shop. The shed was partially rebuilt by BR and closed to steam on 1st May 1967.

Glasgow Central is the location as No 80027 departs at the head of a local service in May 1952. The locomotive was ex-works at Brighton on 4th January 1952 and sent to Polmadie depot where it was to remain until January 1963 when it was reallocated to Eastfield. A final move back to Polmadie took place in December 1964 for its last two years service being withdrawn on 29th November 1966. *J. Flint/J. Harbart (FH976)*

No 80029 is seen at Drybridge, probably on a Kilmarnock-Ayr train, during March 1965. The station was opened on 6th July 1812 by the Kilmarnock & Troon Railway. The Glasgow, Paisley, Kilmarnock & Ayr Railway took over management of the station on 16th July 1846, while its successor, the Glasgow & South Western Railway, took over full ownership in 1899. The station closed to passengers on 3rd March 1969, general freight traffic had ceased 10 years earlier on 2nd November 1959. No 80029 entered service on 22nd January 1952 being allocated to Kittybrewster, where it was to remain until reallocated to Ardrossan in June 1961. A final move to Hurlford took place in the summer of 1962 from where it was withdrawn on 16th December 1965 – destined for breaking up at Faslane. W. *A. C. Smith (7902)*

The crew of No 80030 pose for their photograph to be taken on 26th April 1952. The locomotive was new to traffic on 5th February 1952 being allocated to Ayr, where it was only to remain for a few weeks before being reallocated to Corkerhill on 11th March. A move to Ardrossan took place on 6th October 1958, where it was to remain until 2nd February 1959 when a final move to Corkerhill took place. It was withdrawn on 8th June 1964 and sold to Arnott Young, Troon, for cutting up. *J. Robertson*

Above. No 80031 is seen inside the shed at Brighton alongside classmate No 80010 and Ivatt Class 2MT No 41326. The locomotive was built at the adjacent works, entering traffic on 2nd February 1952. It remained at Brighton until the end of 1963 when it was reallocated to Redhill from where it was withdrawn on 20th September 1964. Sold to the Steel Supply Co, No 80031 was cut up in the goods yard at West Drayton in March 1965. No 80010 entered traffic on 10th July 1951, the first of the class to be completed. Allocated to Tunbridge Wells West when new, it was to be reallocated to Brighton, Three Bridges, back to Tunbridge Wells West with a final move to Brighton from where it was withdrawn on 14th June 1964. Its final journey was to Cashmore's (Newport) for recycling in March 1965. *J. G. Walmsley*

Top right. No 80032 stands at Lymington Pier at the head of the shuttle service from Brockenhurst. Entering service on 4th March 1952 the locomotive was initially allocated to Brighton. It was transferred to Redhill in December 1964 before a final move to Bournemouth in May 1965. That depot provided locomotives for use on the Swanage and Lymington branches that served the coastal communities. The single track branch is around 5.6 miles (9km) in length and was opened, as far as Town station, by the Lymington Railway Co on 12th July 1858. The London & South Western Railway bought out the local company in 1879, and in 1884 the LSWR opened a short extension of the line to Lymington Pier. By 1967, the line was the last steam-hauled branch on BR. The last passenger train ran on Sunday 2nd April 1967 behind LMS Ivatt Class 2 No 41312. No 80032 did not survive to see out steam on the line, being withdrawn on 29th January 1967.

Bottom right. No 80036 was mentioned in *Southern Miscellany* that an enthusiast reported that on 30th June 1952: *New Standard Class '4' 2-6-4Ts for service elsewhere continue to be completed at Brighton and run in from the shed there, including journeys on the evening through train to Tonbridge and back, Nos 80036-9 being recently noted.* Completed on 23rd May 1952 made the journey to the London Midland Region's depot at Watford. It is seen here at Watford Junction on 15th March 1958 in tandem with a Stanier 8F. No 80032 was reallocated to the Southern Region with a move to Ashford in December 1959 with a final move to Exmouth Junction on May 1962/ It was withdrawn from there on 9th November 1964 and sold to Cashmores, Newport, for scrapping. *B. Wilson*

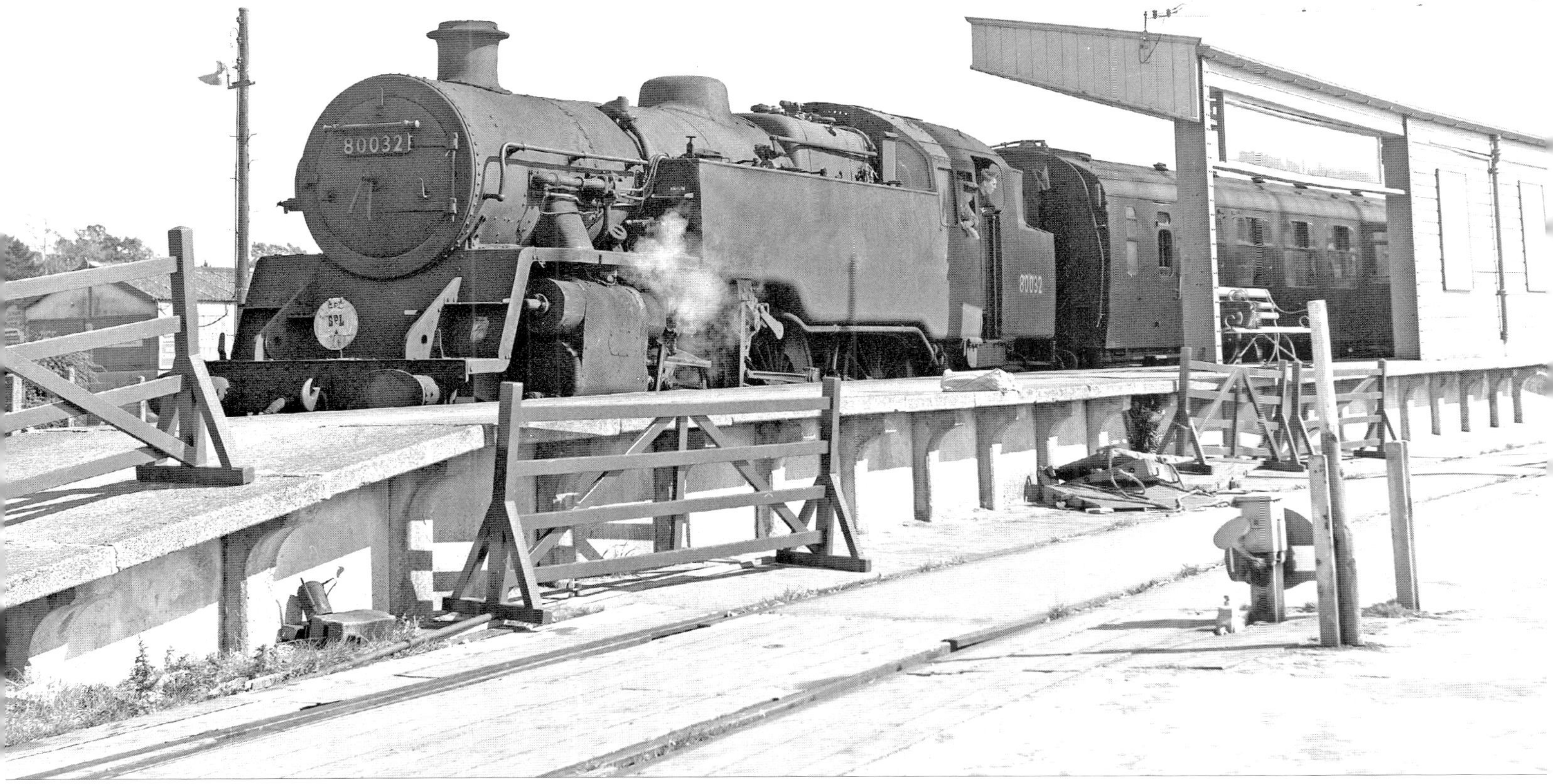
80032
SPL
80032

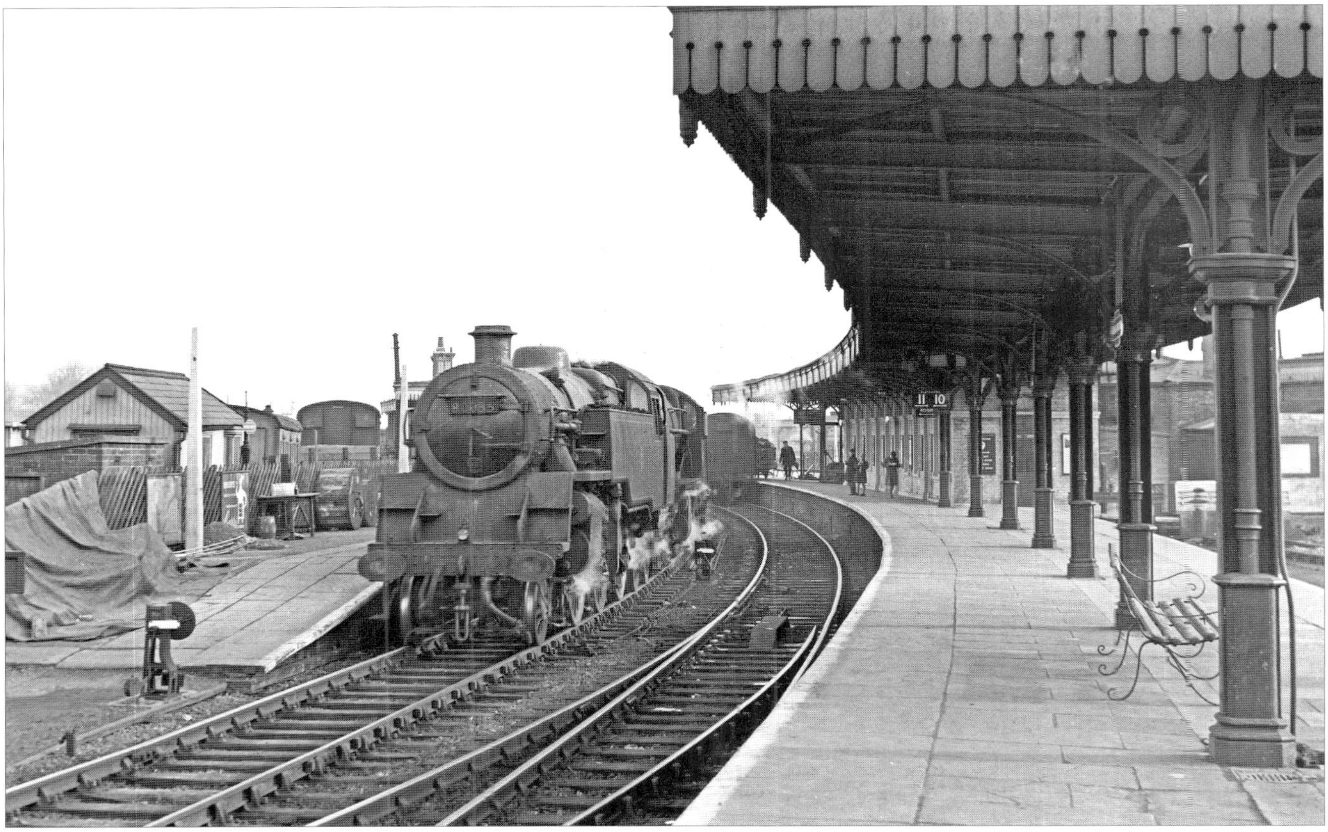
11 10

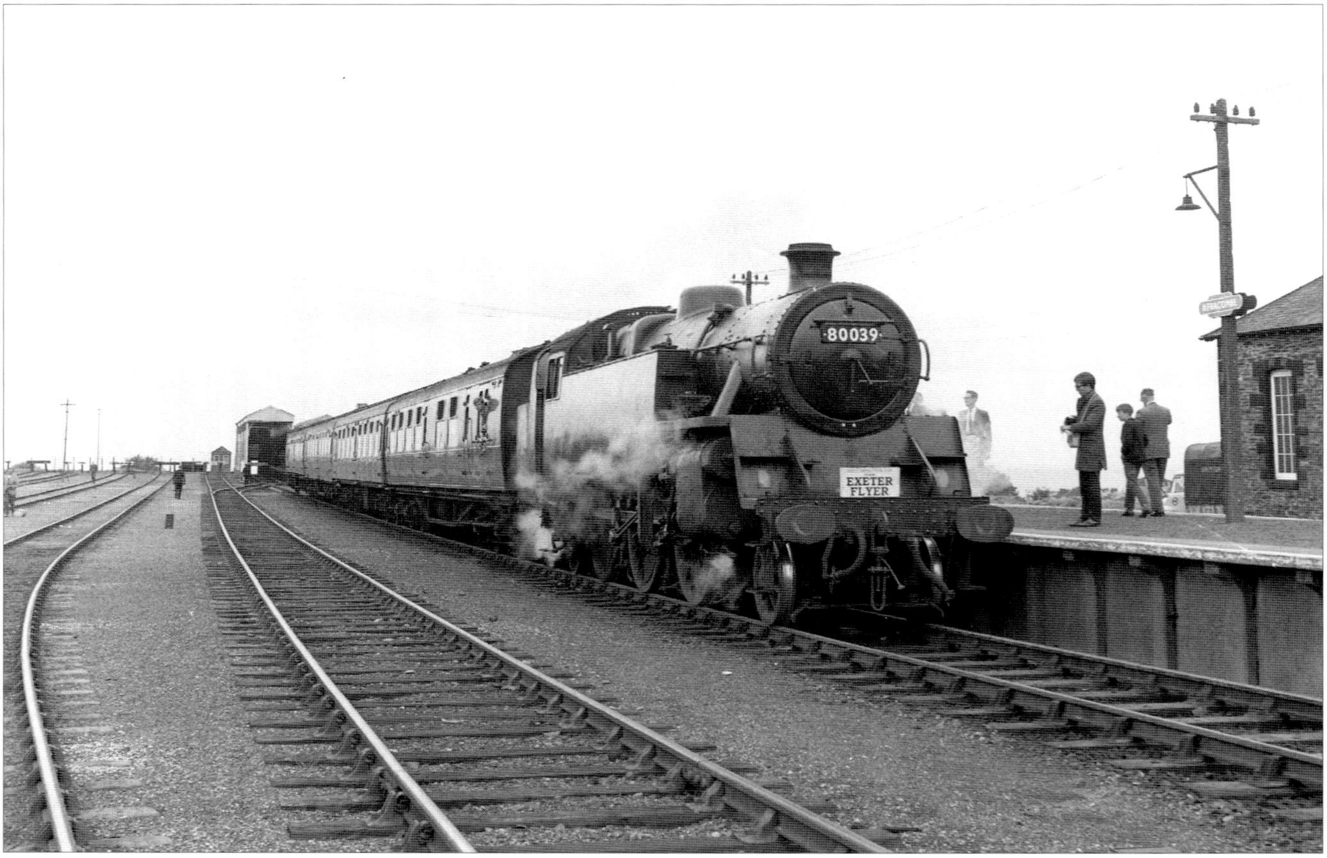
80039
EXETER
FLYER

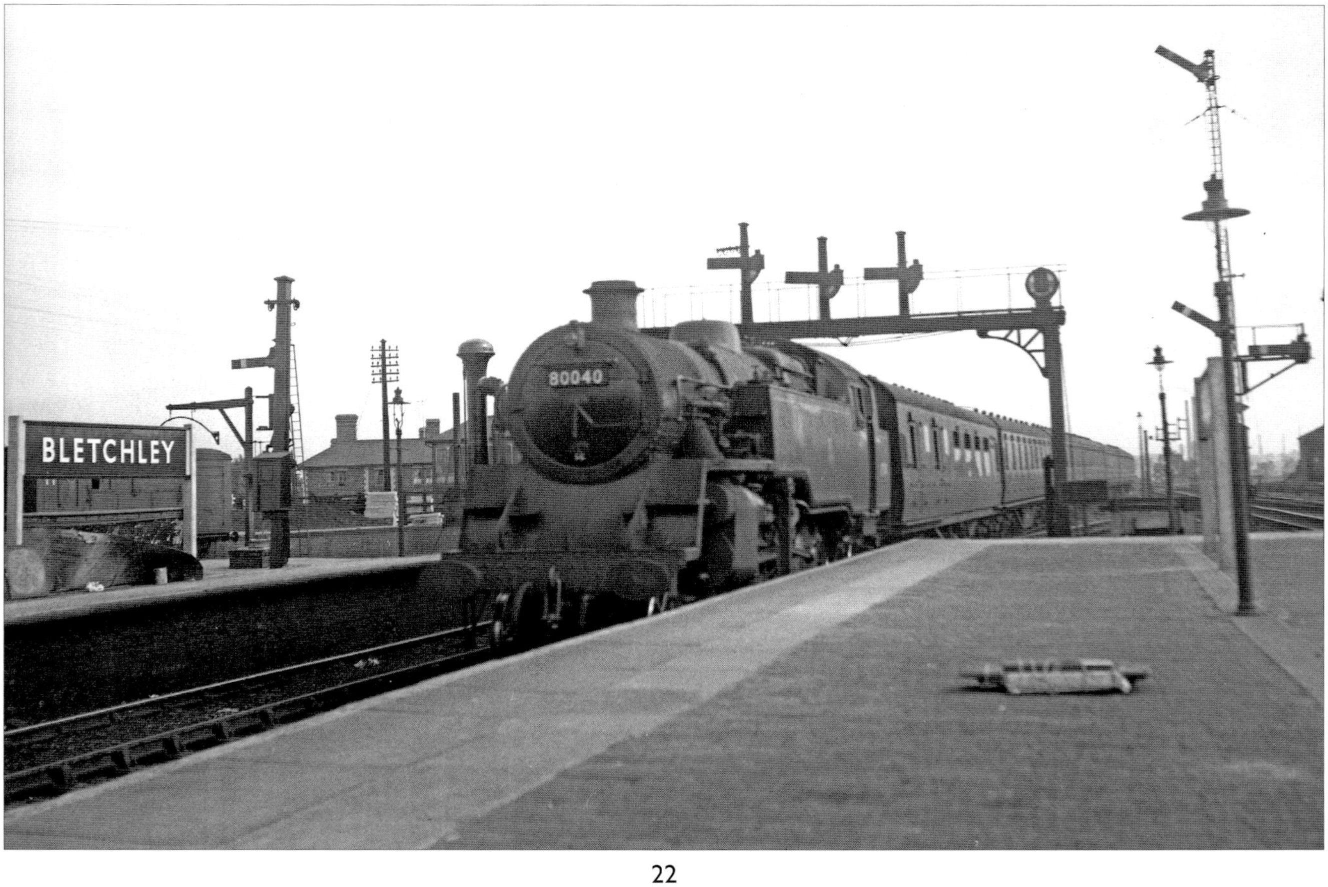
BLETCHLEY
80040

# The Class of 1952

Top left. No 80039 is seen at Ilfracombe on 12th September 1965 at the head of the 'Exeter Flyer' railtour operated by the Southern Counties Touring Society. The locomotive and No 80043 took over the train from Waterloo at Exeter Central, taking it to through Barnstaple where it split. No 80039 then worked a return trip to Ilfracombe, with No 80043 working a round trip to Torrington. No 35022 *Holland-America Line* worked the out and back trip to Waterloo. No 80039 entered traffic on 18th June 1952 at Bletchley – over the next 13 years it saw service at 10 depots, ending its days at the Western Region's Templecombe shed, on the Somerset & Dorset Joint line, where it arrived in the summer of 1965. Withdrawn on 21st January 1966, a couple of months before closure of the S&D, it went to South Wales for demolition. *B. Wadey*

Bottom left. No 80040 had emerged from Brighton Works on 27th June 1952 and was allocated to Bletchley depot. It is seen here arriving at its home station on 26th July 1952 with the 3.47pm service from Watford Junction – already looking in need of a clean! A move to Chester followed in the summer of 1957, with a further transfer to the Southern Region when it arrived at Ashford depot in December 1959. After a move to Tonbridge it arrived at Exmouth Junction during June 1962. It was withdrawn on 6th May 1964, and was one of three of the class to be scrapped in-house' at Crewe locomotive works. *M. N. Bland*

Above. No 80041 heads a North Cornwall line service out of Okehampton, passing the Military Sidings to the west of the station. This was the arrivals point for men, materials and machinery destined for Okehampton Camp – an uphill march for the former! On this occasion the left hand siding is occupied by four-wheel box vans with the siding nearest the running lines hosting hopper wagons containing ballast from the nearby Meldon Quarry. The locomotive was delivered new to Bletchley in July 1952 before a move south to Ashford in December 1959. As with No 80039, after a move to Tonbridge it arrived at Exmouth Junction during June 1962, staying until early summer 1965 when it moved to the county of Somerset. It was withdrawn from Templecombe depot when the S&D line closed on 7th March 1966.

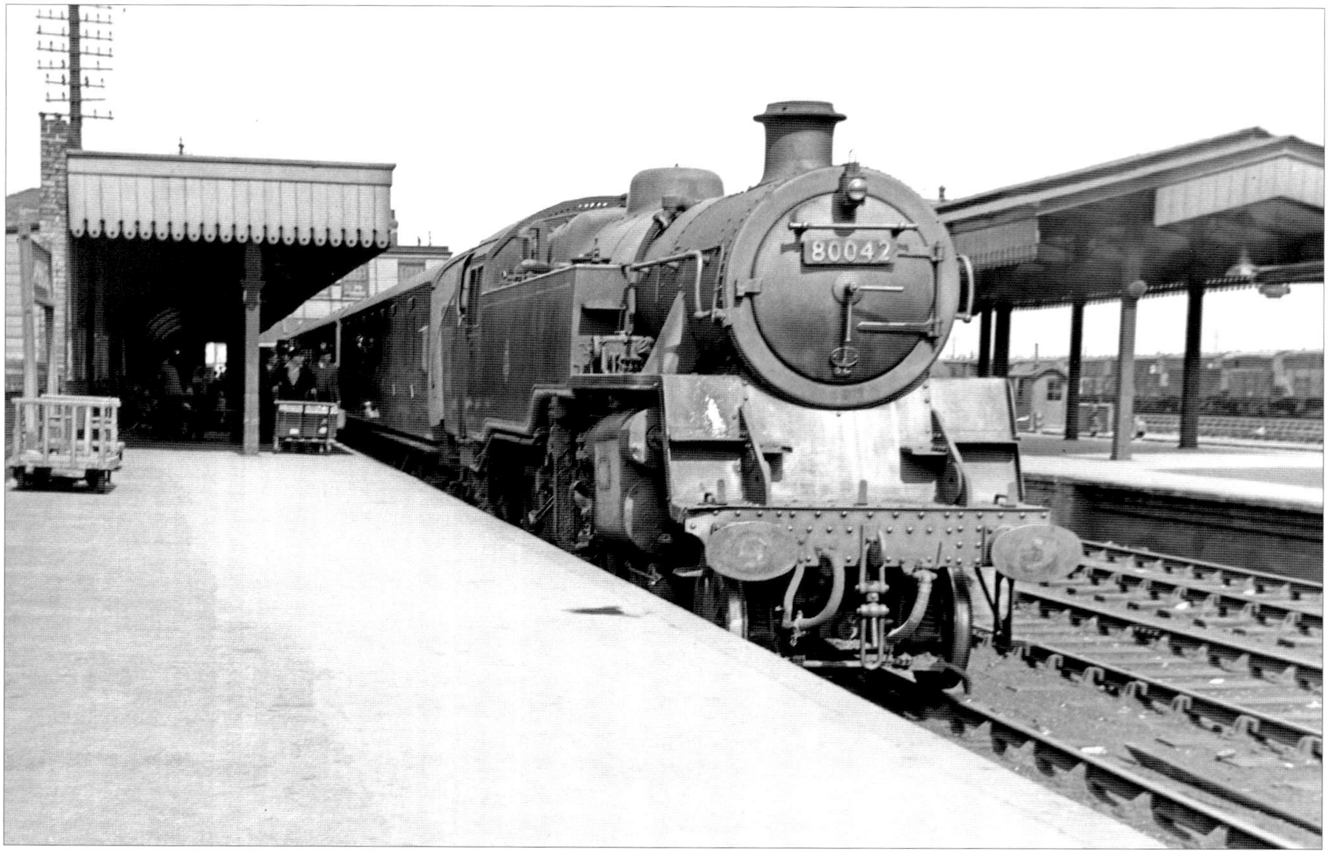
80042

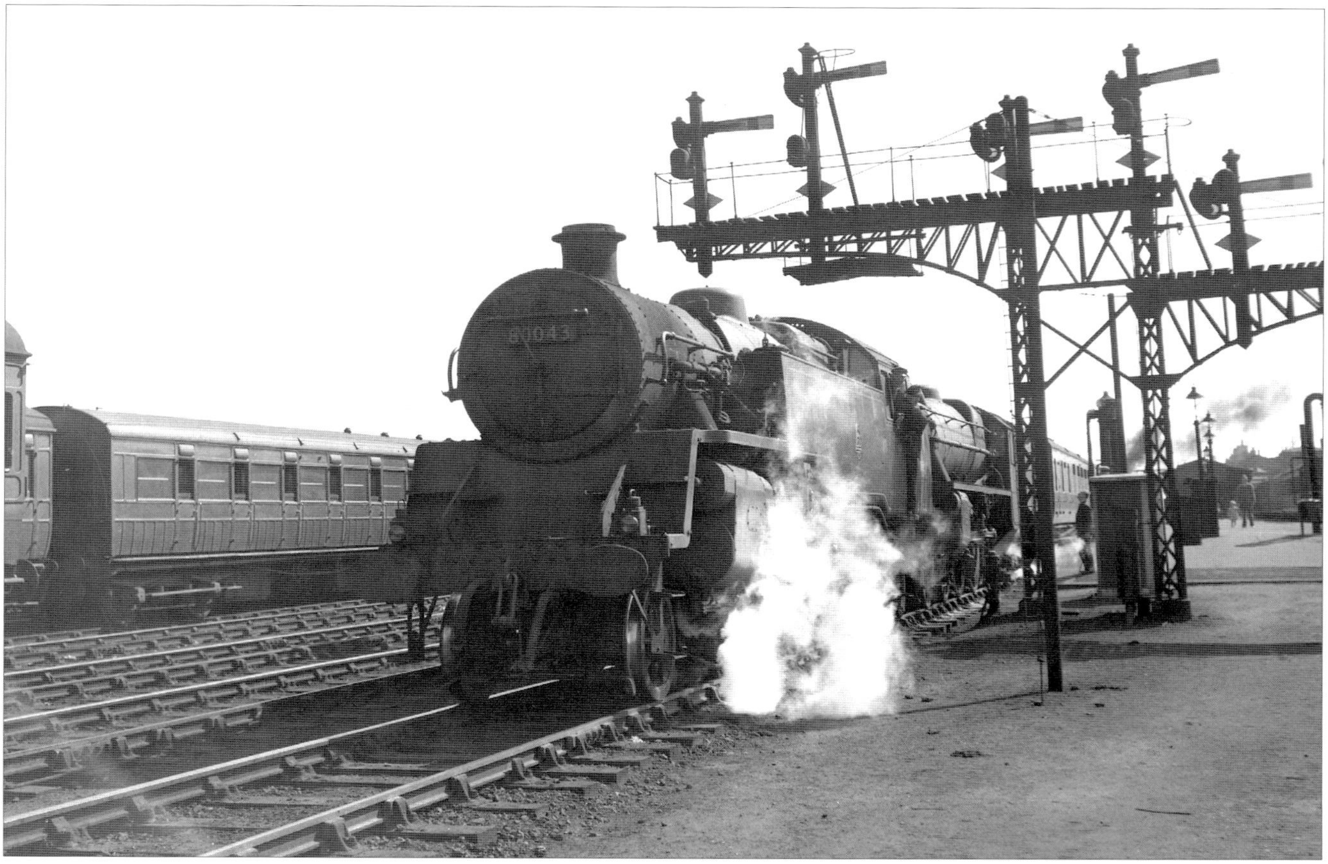

# The Class of 1952

Top left. Sitting in the bay platforms at the southern end of Cambridge station, No 80042 waits to take a local service to Bedford via the former London & North Western Railway's (LNWR) route via Sandy. The locomotive was initially allocated to Bletchley where it had arrived following completion at Brighton Works on 15th July 1952. A transfer to the Southern Region occurred in December 1959 when it arrived at Ashford, taking the usual route to Exmouth Junction via Tonbridge. It arrived in the West Country in June 1962, where it survived until withdrawal on 6th February 1965 when it went to Hayes scrapyard at Bridgend – the only Standard Four tank to be broken up there. *E. Sawford (ES1114)*

Bottom left. Cambridge is the location on 10th May 1953 as Nos 80043 and 44771 double head a troop special from Bletchley to Weybourne, consisting of Western Region coaching stock. Weybourne camp originally started out as a temporary summer residence for the Anti Aircraft Division of the Territorial Army in 1935. Located alongside the cliffs at Weybourne to the north-west of the village, at first the majority of the camp consisted of wooden and tented structures, although in 1937 it was decided to make the camp permanent and more fixed structures and defences were erected. During World War 2 the camp was surrounded by a perimeter anti tank ditch and defended by a system of gun emplacements and barbed wire. The interior of the camp consisted of groups of Nissen huts and barracks and other military buildings. The camp closed in 1959 and the site is now part of the Muckleburgh Collection, which is a military museum that was opened to the public in 1988 and is the largest privately owned military museum in the United Kingdom. No 80043 was allocated new to Bletchley in July 1952, remaining there until December 1959 when it was reallocated to Dover Marine. A move to Ashford was followed by stays at Tonbridge and Exmouth Junction before arriving at Templecombe in late summer 1964. It was withdrawn when the S&D Joint closed on 7th March 1966. *M. N. Bland*

Below. No 80045 skirts the promenade at Saltcoats with a down passenger extra, reporting No 68, on 20th July 1964. New to traffic on 4th September 1952 the locomotive was allocated to Bedford, where it stayed until January 1955 when it moved to Kentish Town. September 1956 saw a move north to Chester, before reallocation to the Scottish Region at Corkerhill, followed by Beattock and Polmadie. It arrived at the latter on 21st November 1965 from where it was withdrawn on 1st May 1967 and consigned to Campbells, Airdrie, for scrapping. *W. A. C. Smith (WS7561)*

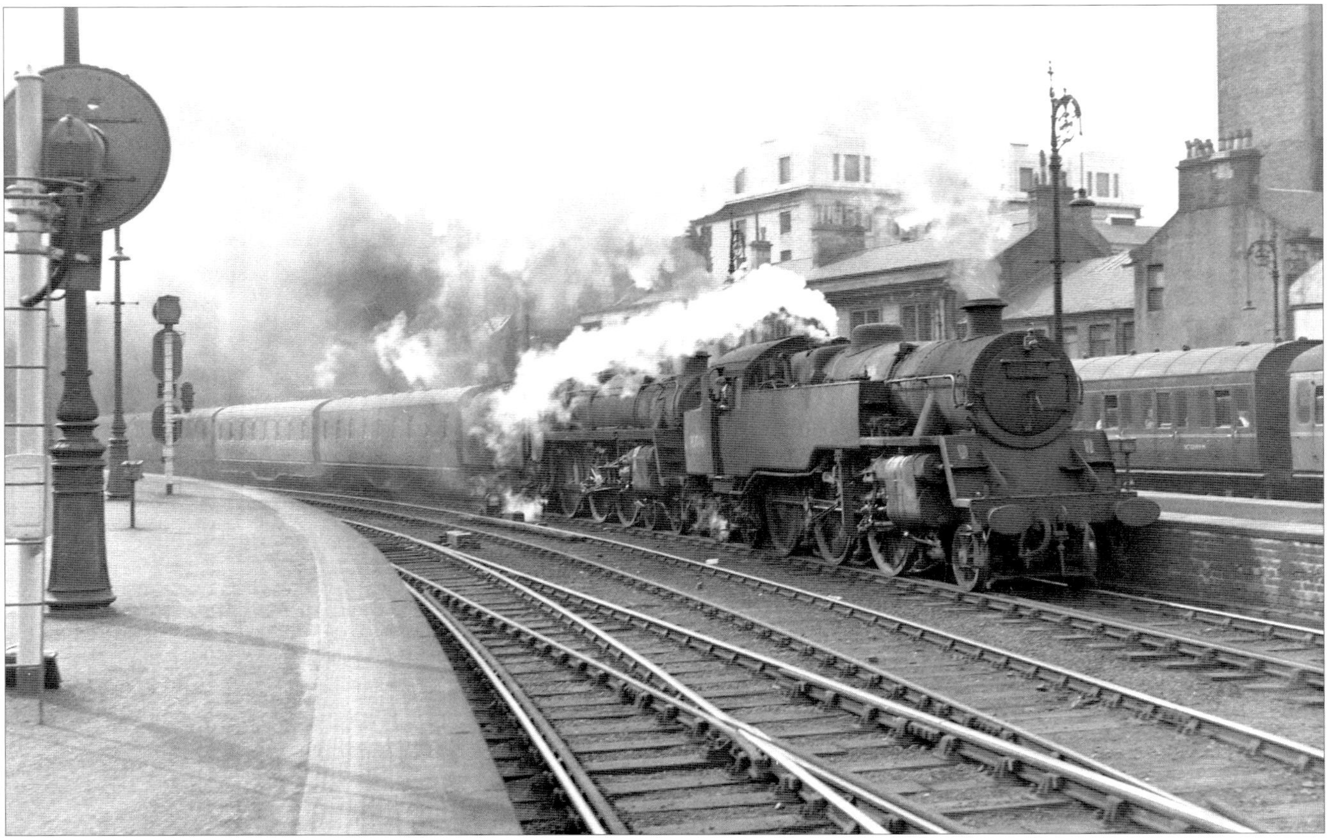

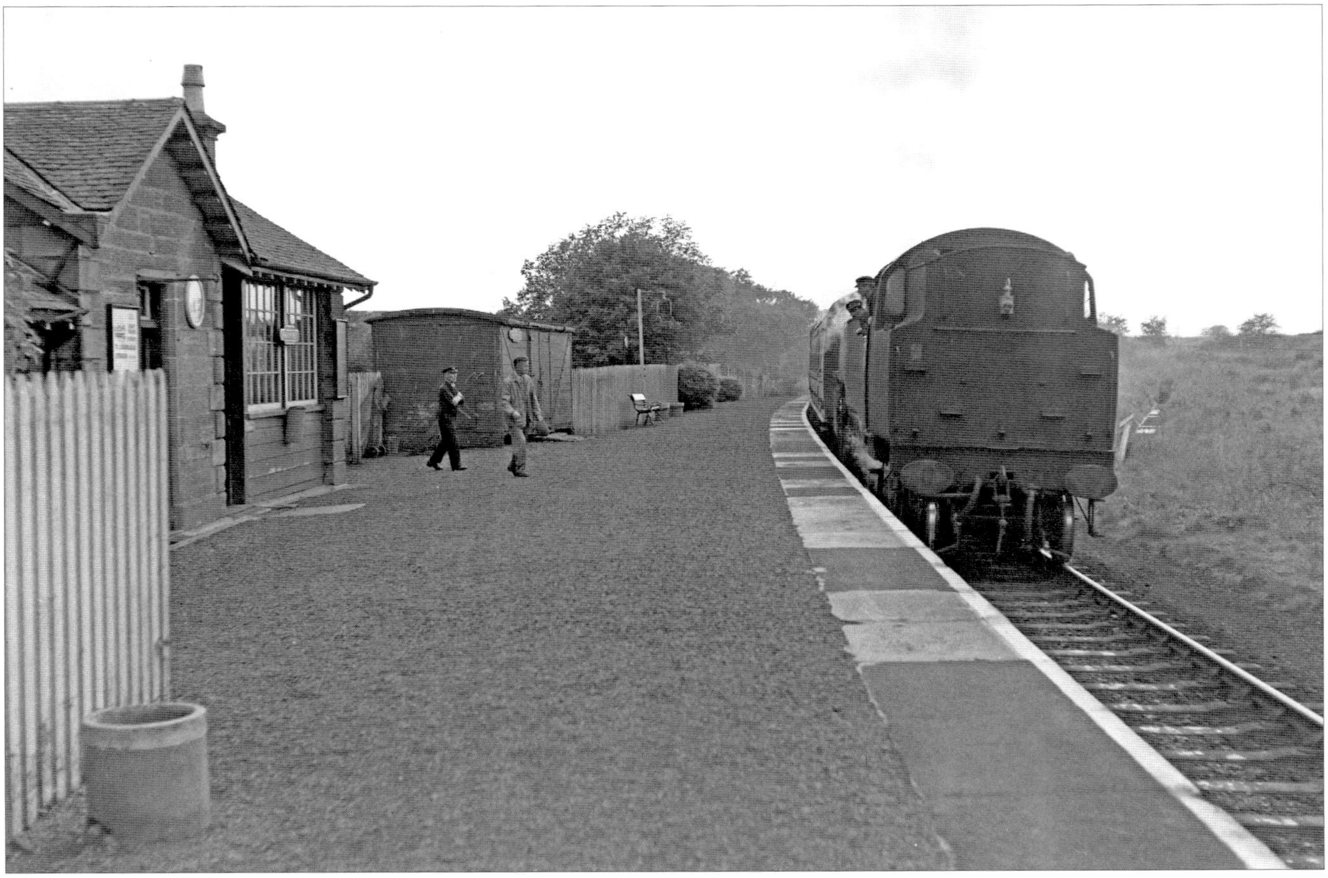

# The Class of 1952

Top left. No 80046 double-heads an unidentified Standard 5 at Glasgow St Enoch with an Ayrshire coast semi-fast service on 25th June 1964. The locomotive was new to Bedford on 17th September 1952 and moved north to Bury in January 1955. Reallocated via Newton Heath and Blackpool Central, it arrived at Corkerhill on 4th February 1960. It was withdrawn on 1st May 1967, being disposed of by Campbells, Airdrie, the following August. Glasgow St Enoch was originally built for the cross-city City of Glasgow Union Railway but became the main terminus, and headquarters, of the Glasgow & South Western Railway. The original 1876 built station consisted of six platforms, ultimately being extended to 12 by 1902. After closure to passengers on 27th June 1966 it remained open as a parcels depot for a short period before becoming a car park. The building was demolished in 1977 with the stonework being used to fill the former Queens Dock. *W. A. C. Smith (WS8514)*

Bottom left. Running bunker first No 80047 stands at Hairmyres with the 5.08pm St Enoch-East Kilbride service on 10th September 1964. The Caledonian Railway opened the station on 1st September 1868 with just one platform on the single line between Busby and East Kilbride. Although the station closed to general goods traffic on 5th July 1965 sidings remained until the 1970s to serve the Radio Times factory in the adjacent College Milton. No 80047 was new to Bedford on 30th September 1952 with a move to Kentish town the following January. In September 1956 it moved north to Chester, a final move to Corkerhill occurred in March 1960. Withdrawn on 24th August 1966 it was despatched by Shipbreaking Industries cutters at Faslane that October. *W. A. C. Smith (WS7756)*

Below. At the head of a local two-coach service No 80049 waits for departure time at Kilmarnock on 16th July 1963. The first station was opened by the Kilmarnock & Troon Railway on 6th July 1812, one of the earliest in Scotland. It was replaced by the Glasgow, Paisley, Kilmarnock & Ayr Railway on 4th April 1843 with the opening of its main line from Dalry. The third and current station was opened on 20th July 1846 by the Glasgow Barrhead, Paisley, Kilmarnock & Ayr Railway – this was connected to Ardrossan via Irvine two years later and to Carlisle via Dumfries & Gretna Junction in 1850. The current route to Glasgow (via Barrhead) – the Glasgow & Kilmarnock Joint Railway – was completed in 1873 jointly by the G&SWR and Caledonian Railway. No 80049 was new to Newton Heath 24th October 1952, moving to Chester on 15th September 1956. Transfer to the Scottish Region came on 5th May 1960 with a transfer to Corkerhill. Withdrawn on 8th June 1964 it was sold to Arnott Young, Troon, for recycling. *J. L. Stevenson*

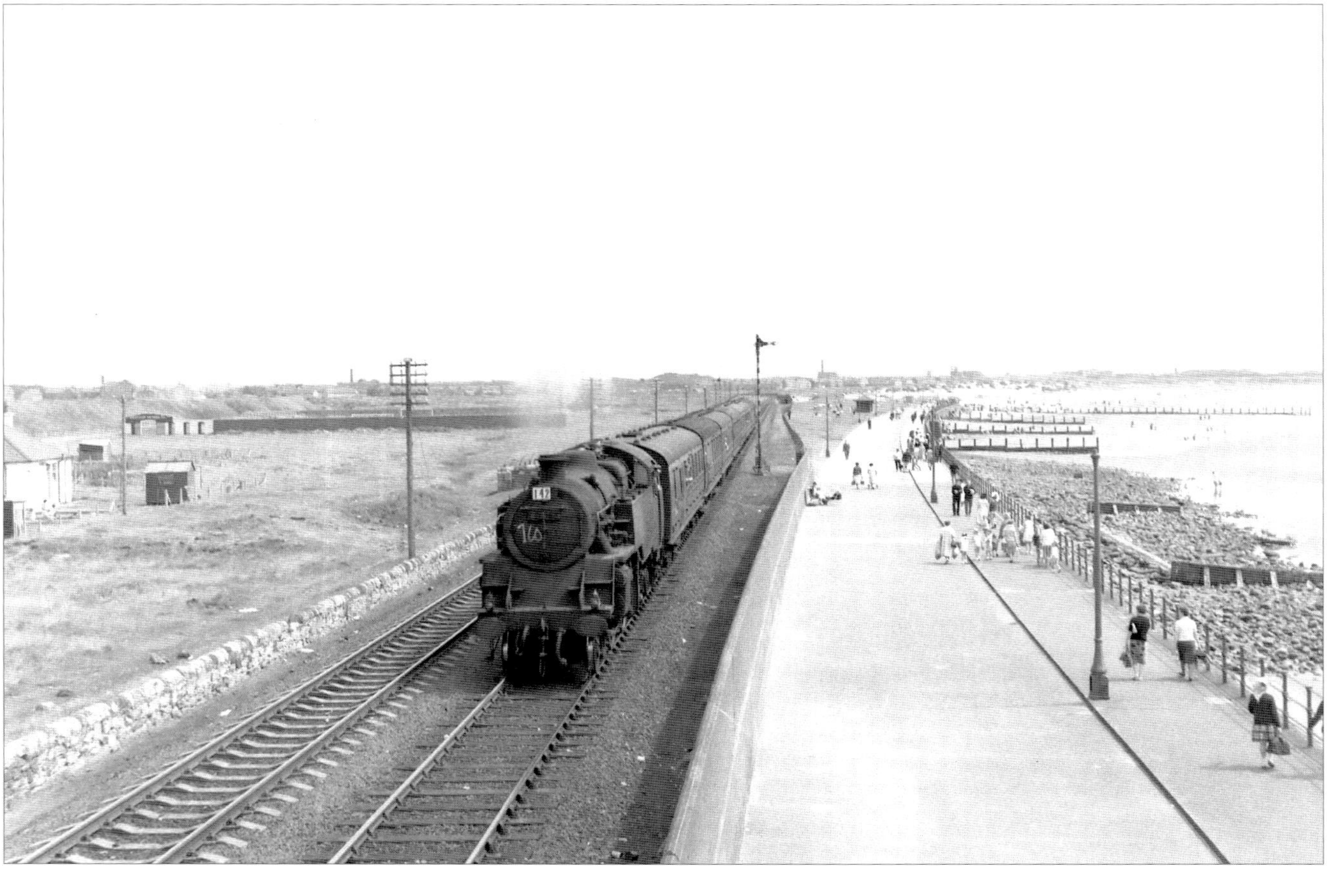

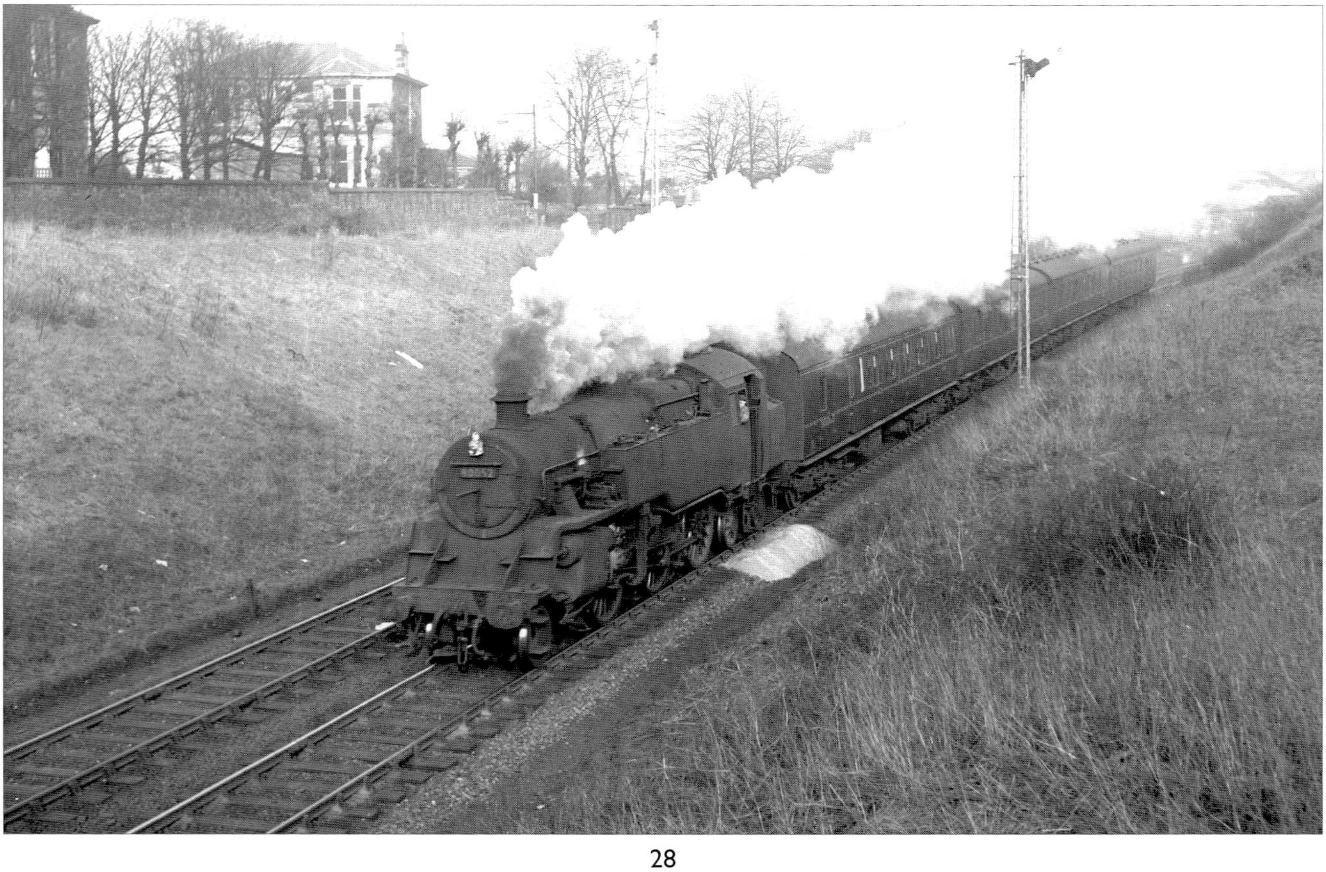

Top left. No 80051 is seen passing Saltcoats promenade on a Glasgow Fair Monday excursion, train No 42, to Largs on 20th July 1964. In the 1500s, King James V dipped into his own pocket to establish the salt panning industry in Saltcoats, from which the town takes its name. The small harbour dates from the late 17th century with later alterations, and at low tide fossilised trees can be seen on the harbour floor. It was in Saltcoats in 1793 that Betsy Miller, the first woman ever to have become a registered ship's captain, was born. No 80051 was allocated new to Newton Heath on 21st November 1952 from Brighton Works. A move to Chester followed in September 1956 – probably travelling in company with Nos 80048-53 that were reallocated at the same time. Nos 80044 to 80053 were among the 19 members of the class, the last in use in the London Midland Region, transferred to Scotland in the spring of 1960 in exchange for a similar number of Fairburn 2-6-4Ts. A final move to Corkerhill took place on 7th May 1960 from where it was withdrawn on 17th August 1966 before being dragged to Shipbreaking Industries cutters at Faslane that October, in company of several others of the class. *W. A. C. Smith (WS7586)*

Bottom left. The 1.15pm Glasgow St Enoch-Kilmarnock service is seen at Bellahouston, on the Glasgow & South Western Railway's Paisley Canal line, behind No 80052 on 15th February 1964. This was the last of the class to be delivered in 1952, being allocated to Newton Heath on 5th December. It was reallocated to Chester, along with other members of the class, on 15th September 1956 and on to Corkerhill in March 1960. To the right of the image, hidden in the murk, can be seen the cranes of Clutha Iron Works that was in operation from 1872 until 1970 – the site of the works is now an industrial estate. No 80052 had a much shorter life, surviving in traffic for less than 12 years; being withdrawn on 7th July 1964 it was sold to Motherwell Machinery & Scrap, Wishaw, for recycling. *J. L. Stevenson*

Above. The first of the Class 3MTs to be completed left Swindon Works on 4th April 1952 bound for Tyseley depot. No 82000 left Birmingham for South Wales in July 1953 when it was reallocated to Barry in South Wales. Relocated to Treherbert in September the same year it arrived at Shrewsbury depot in November 1955. It is seen here on 13th August 1958 with the 6.40am Shrewsbury to Bridgnorth service. January 1959 saw a move to Wrexham Rhosddu for a short stay as it left for Machynlleth the following January. A move to the Scottish Region occurred in March 1965 when it was reallocated to Polmadie depot. It was withdrawn on 10th December 1966 and despatched to South Wales for recycling. *R. Wilson*

Above. It was not the fault of the locomotives, but the Class 3MTs were to chase work – only those initially allocated to the Southern Region had any stability. No 82001 was delivered new to Tyseley and over its 13 year life span was to work out of 11 different depots. It arrived at Templecombe S&D depot on 23rd April 1961 where it is seen in the company of Nos 3215 and 82002 on 7th July 1962 – three months before a move to Hereford. The two road shed seen behind No 82001 dated from 1951 when the original was totally rebuilt. The depot closed with the line on 7th March 1966. Following its stay at Hereford, No 82001 arrived at Exmouth Junction in April 1963, staying until the end of the year when a move to Taunton took place. It returned to Bristol Barrow Road for a second time during June 1964. It was withdrawn on 31st December 1965 and sent to South Wales for demolition. *S. Summerson (SUM776)*

Top right. No 82002 is seen on shed at Chester West depot in company with an unidentified pannier tank. Note the deflectors in front of the mechanical lubricators on the front of the running plates. The locomotive was new to Tyseley on 28th April 1952. It had arrived in July 1958 via Barry, Newton Abbot, Treherbert and Bristol Bath Road. It left Chester West in early April 1960, moving across town to the former Midland Railway depot, before leaving for Templecombe in April 1961, then onto Shrewsbury and Hereford before arriving at Exmouth Junction in April 1963. It was withdrawn from the latter depot on 7th February 1964 and cut up at Eastleigh Works on Saturday 30th May the same year. *R. Bruce*

Bottom right. A nice side on view of No 82003 taking water whilst waiting for departure time at Leamington Spa General during August 1952. The first station at the site, under the name Leamington, was opened by the Great Western Railway on its new main line between Birmingham, Oxford and London in 1852. It was later renamed Leamington Spa in 1913, and again in 1950 when it acquired the suffix General; only to lose it 18 years later. No 82003 had arrived at Tyseley depot following completion at Swindon Works on 2nd May 1952. A move to Barry followed in September 1953, and ultimately following stays at both Chester depots and all three of Bristol's it eventually reached Patricroft, via two more Welsh depots, in March 1965. It was withdrawn on 10th December 1966, returning to South Wales for recycling. *J. Flint/J. Harbart (FH971)*

Above. Following its release from Swindon Works on 14th May 1952, No 82004 was sent to Tyseley where it stayed until September 1953 when it was reallocated to Barry. It returned to England in April 1955 when it arrived at Newton Abbot where it stayed until the summer of 1956 when it was sent to Wellington, Shropshire, where it is seen the following year. No 82004 was reallocated in 1959 with stints at Shrewsbury, back to Wellington, and a final move south to Bath Green Park from where it was withdrawn on 1st October 1965. The shed at Wellington was opened by the GWR in 1876 and included coaling facilities and a turntable – the latter was removed in the 1930s. The three-road shed was closed by BR on 10th August 1964 and subsequently demolished to create a car park.
*N. E. Stead collection (NS207830)*

Top right. No 82005 was allocated to Tyseley on its release from Swindon Works on 16th May 1952. It moved around a fair bit with stays at Barry (from September 1953), Newton Abbot, Treherbert, Bristol Bath Road, both Chester depots, Shrewsbury and Machynlleth. It departed the latter during April 1961 and headed south to London's Nine Elms depot. It is seen near the end of its life serving BR at Clapham Junction, in charge of an empty stock working from Waterloo on 13th July 1965 – it would survive until 19th September that year when it was withdrawn from traffic. The construction above the rear carriage is Clapham Junction 'A' signal box. Opened in 1907 it gained notoriety on 10th May 1965 when it partially collapsed, closing all lines into Waterloo station during the height of the rush hour. One of the London side supporting girders had been consumed by rust, dropping the structure by 3ft 6in. Within a short period of time the structure was jacked up and reopened. *B. Wadey*

Bottom right. No 82006 is seen near its birthplace, having worked in with the 12.38pm service from Andover. It had been reallocated to Bristol Bath Road depot a few days earlier, from its previous depot of Wellington, Shropshire. Leaving the nearby Works on 27th May 1952 it had followed its predecessors to Tyseley, before making the usual move to Barry. Bristol Bath Road closed its doors to steam on 12th September 1960 resulting in a move across the city to Barrow Road, remaining there until a move to Machynlleth the following February. It was to move to the Southern Region at Nine Elms at the end of April 1965, remaining in service until 18th September 1966. *D. Idle*

Fresh out of the box at Swindon on 5th June 1952, No 82008 made the usual trip to Tyseley, staying until September 1953 when it made its way to Barry. It then returned to England with stays at Kidderminster and Worcester before heading back to Wales, arriving at Machynlleth during the early summer of 1961. It is seen here at Barmouth, on the Cambrian Coast line, on 7th September 1961, heading a south-bound service. It moved via Neyland to its final depot at Taunton where it arrived during October 1961, being withdrawn just over three years later on 7th February 1964. It was sent to Eastleigh for disposal and was noted intact in the scrapyard on 31st May 1964, it was broken up the following month. S. Summerson (SUM669)

No 82010 is seen at the fairly rudimentary Polsloe Bridge station on 11th May 1958. The station was opened by the London & South Western Railway in 1907 to serve the eastern suburbs of Exeter. It was situated just a short distance along the branch to Exmouth and was convenient for Exmouth Junction engine shed that was on the opposite side of the main line. The platforms were rebuilt in 1927 using concrete components cast at the concrete workshop that had been established at Exmouth Junction. On 4th February 1973 the branch was singled and the down platform taken out of use. No 82010 left Swindon Works on 25th June 1952 and was allocated to Exmouth Junction depot where it would remain for over 10 years before being transferred east to Eastleigh. Two months later, in November 1962, it would move to Nine Elms and would serve for another two years or so before withdrawal on 24th April 1965 with demolition taking place in South Wales by Birds, Morriston. *R. Wilson*

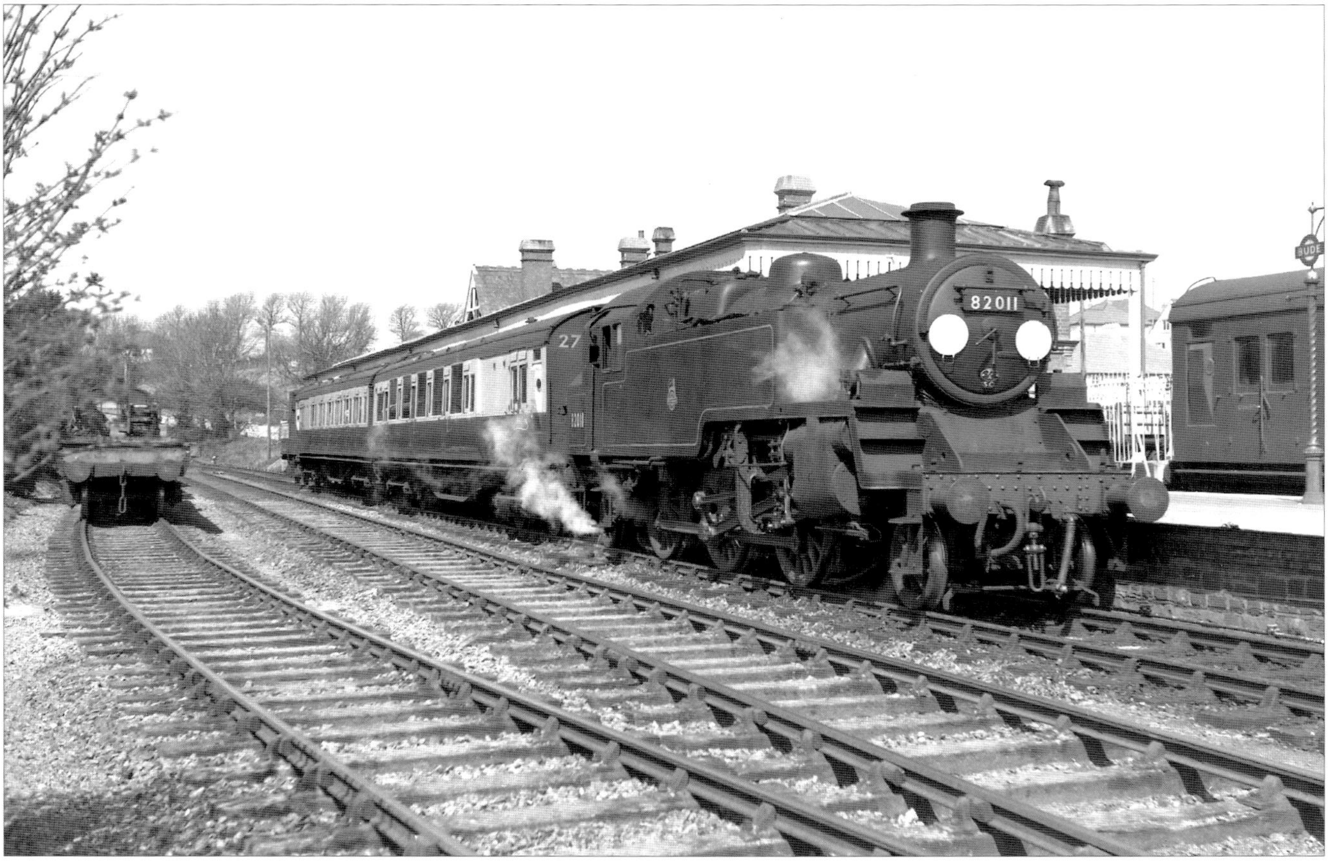
82011
27

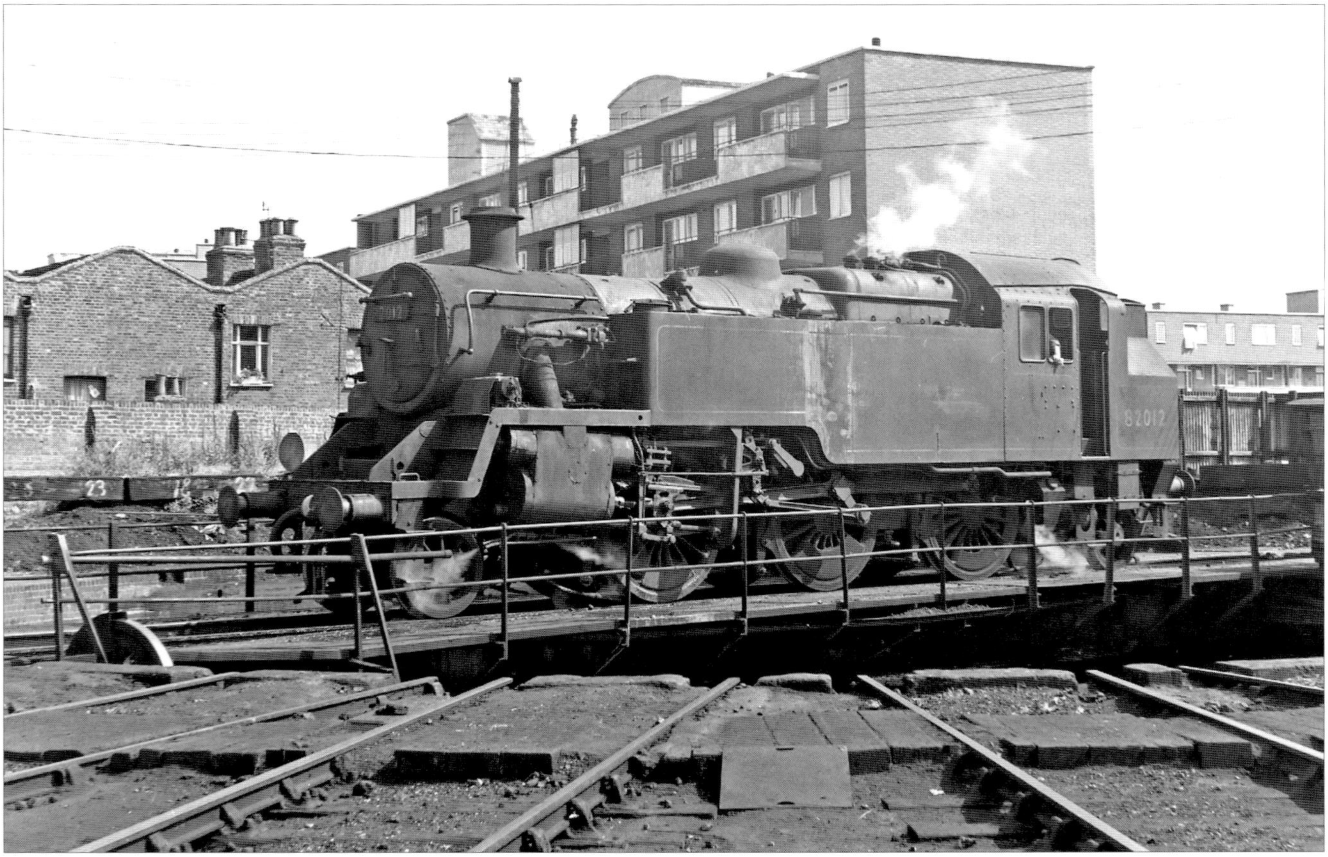
82012

# The Class of 1952

Top left. On 17th April 1955, No 82011 is a fair distance from its home depot of Exmouth Junction, having made the trip to Bude, although it may well have been sub-shedded at Okehampton at the time. It is seen here about to depart with the 2.32pm Bude to Okehampton service. The opening of Bude station on 11th August 1898 marked the completion of the LSWR's branch line from Okehampton that had taken 19 years and four Acts of Parliament. The original line had been authorised as far as Holsworthy where a station was opened on 20th January 1879. The station closed to passenger traffic on 3rd October 1966, freight traffic had ceased earlier on 7th September 1964, leaving residents of Bude and the surrounding area with Okehampton station, some 30 miles (48 km) away, as their nearest connection to the railway. This increased to 33 miles (53 km) in January 1972 when Okehampton itself closed, leaving Barnstaple as the nearest railhead. No 82011 entered service from Swindon Works on 1st July 1952 and headed south to Exmouth Junction, staying there until it moved to Eastleigh in September 1962. A final move to Nine Elms, probably in company with No 82010, took place. It was withdrawn two years later on 30th August 1964, making its final trip to Cohens, Kettering where it was broken up that November. *J. Flint/J. Harbart (FH902)*

Bottom left. The first depot at Nine Elms was opened by the London & Southampton Railway on 21st May 1838. The original three-road building was extended and eventually consisted of two buildings – a 15 road dead-ended building (the 'old shed') and a 10 road dead-ended version (the 'new shed'). Following closure the site was cleared and is now occupied by a fruit and vegetable market. No 82012 entered traffic on 4th July 1952 and arrived at Nine Elms in December 1962 following stints at Exmouth Junction and Eastleigh. It survived in traffic until 31st May 1964; it was broken up by Cohens, Kettering, that November. No doubt the local non-railway employee residents were grateful when time was called on Southern Region steam traction in the summer of 1967. *J. Flint/J. Harbart (FH2934)*

Below. A scene that would cause today's Health & Safety Officer to suffer a heart attack as young enthusasts climb over No 82014 and adjacent 350hp diesel shunter (probably No 15235). It is seen at Eastleigh Works on 7th April 1957, the locomotive had arrived from the adjacent depot for a Light Intermediate overhaul on 18th January 1957 that was completed on 9th February. The London & South Western Railway opened a carriage and wagon works at Eastleigh in 1891. In 1903, the Chief Mechanical Engineer, Dugald Drummond, oversaw the construction of a large motive power depot in the town; replacing the existing maintenance and repair shops at Northam, Southampton. In January 1910, locomotive building was likewise transferred to the new workshops at Eastleigh from Nine Elms in London. In 1950, following nationalisation, new steam locomotive building ceased at Eastleigh. However the works were kept fully occupied between 1956 and 1961 in rebuilding over 90 of the Bulleid Pacifics. Thereafter the works gradually changed over to steam and diesel repairs. In 1962, the works was again reorganised with the carriage works site being sold, and carriage and electric multiple unit repairs transferred to the main locomotive works. Ultimately the works closed in March 2006 due to lack of work but was reopened under 'new management'. No 82014 entered traffic on 1st August 1952 and was allocated to Eastleigh depot from new. In December 1962 it was transferred to Nine Elms from where it was withdrawn on 31st May 1964 – disposal was by Cohens, Kettering, that November.

# The Class of 1952

Top left. No 82015 left Swindon Works on 13th August 1952 and headed to Eastleigh where it would remain for the next 10 years. It is seen here arriving at Southampton Central with the 1.35pm Northam-Fawley refinery oil tanks on 19th September 1961. Note the barrier wagons between the locomotive and the leading tank wagon. The construction of the oil refinery in 1920/21 by Anglo Gulf West Indies Petroleum Corporation Ltd took place and, during the building, potential means of transporting products from the refinery were considered. Initially a pipeline to a storage depot at Lyndhurst Road or Beaulieu Road stations was proposed but this was dropped in favour of a rail connection between the refinery and the London & South Western Railway (L&SWR) at Totton. Early in 1921, The Totton, Hythe & Fawley Light Railway Co was inaugurated. Its aim was 'To construct, maintain and work a railway, 9 miles and one furlong in length from Totton to the refinery'. The line opened to passenger traffic on 20th July 1925. Closure to passenger services came on 14th February 1966. At this time the three stations remained open for goods traffic but final closure to all traffic came on 2nd January 1967. The line would remain open for traffic to Marchwood MoD depot and the oil refinery. On 1st September 2016 the final delivery of crude oil by rail arrived at Fawley following the refinery's decision to receive all crude oil by sea. However, despite the well publicised 'last train' the previously scheduled 09.21 Fawley Esso-Eastleigh departed on time on 13th September – the down working, 07.39 ex-Eastleigh, was cancelled and a light locomotive sent to Fawley instead. This working would have probably been to collect wagons still standing at Fawley. Following its stint at Eastleigh No 82015 was transferred to Guildford during December 1962, remaining there until March 1963 when it moved to Nine Elms. It would be withdrawn on 6th December 1964, destined for Cashmore's scrapyard at Newport, South Wales. *B. Wadey*

Bottom left. No 82016 stands in the Works yard at Swindon shortly after completion in August 1952. It was allocated to Eastleigh on the 21st where it remained until November 1962 when it was transferred to Guildford. A final move to Nine Elms took place in March 1964 from where it was withdrawn on 25th April 1965. Alongside is GWR Collett 2-6-2T No 5537 partially stripped ready to enter the works. New in July 1928, it was withdrawn on 17th August 1962 after 34 years in traffic. *J. T. Clewley*

Below. No 82017 stands outside the engine shed at Exmouth Junction in the mid-1950s. It had arrived from Swindon Works in August 1952 and would remain at the depot until reallocated to Eastleigh in September 1962. Two months later it would head east to Nine Elms from where it would be withdrawn on 25th April 1965, being consigned to Birds at Morriston for scrapping. The original Exmouth Junction shed was opened by the L&SWR on 3rd November 1887. The main shed was built in corrugated iron and covered 11 tracks. A 55ft (17m) turntable was situated behind the shed and a range of other facilities were provided including a dormitory for engine crews and a wagon repair workshop. Work on a replacement shed started in the summer of 1924. The main shed was now of concrete construction 270ft (82m) long and 235ft (72m) wide across 13 tracks. A new 65ft (20m) turntable was provided, and a mechanical coaling tower with a capacity of 300 tons built from concrete replaced the old wooden coaling platform. The first seven tracks were brought into use in 1926 and the final work was completed in 1929. At its height in excess of 400 staff were based at the depot, including 240 locomotive crew. The turntable was replaced again in 1947, this time by a 70ft (21m) example. Part of the Southern Region from 1948, it was coded 72A. In 1963 it was transferred to the Western Region and the code was changed to 83D. It was closed to steam on 1st June 1965 and the staff transferred elsewhere in 1966, although a few diesels were stabled there until final closure on 6th March 1967.

Nos 82018 and 82023 wait for departure time at Sidmouth with a service bound for London Waterloo on 20th July 1957. The two Class 3MTs would take the train to Sidmouth Junction where it would be combined with a service from Exeter Central, probably hauled by a Bulleid Pacific. No 82018 carries the 1956 totem, no doubt dating from a General overhaul at Eastleigh earlier in the year. New to traffic on 8th September 1952 it would follow the usual route, being transferred to Eastleigh in September 1962, and Nine Elms two months later. It would be withdrawn on 10th July 1966 and demolished by Buttigiegs of Newport. The Sidmouth Railway got its Act of Parliament on 29th June 1871. It was to be constructed under the arrangements for a Light Railway, and an agreement was made with the L&SWR to operate the line – it opened on Monday 6th July 1874. Through coaches were discontinued in 1964 except on summer Saturdays, as the local trains were diesel multiple units. The line closed to passenger traffic on 6th March 1967 and to freight on 8th May that year. *J. Harrold (H109)*

Missing its smoke box door number plate, the end of steam is approaching on the Southern as No 82019 departs Waterloo with an empty stock train to the carriage sidings at Clapham Junction on 15th April 1967. New from Swindon Works on 29th September 1952, it would take the route of its predecessors to arrive at Nine Elms in November 1962. It would survive until the end of Southern steam on 9th July 1967. *B. Wadey*

# The Class of 1953

Already looking work weary, No 80059 has been in traffic for a little under three months following completion at Brighton Works on 3rd March 1953. It is seen here on the 10.55am Bedford-St Pancras service leaving Elstree Tunnel on 28th May that year. Initially allocated to Kentish Town, the locomotive was reallocated to Chester on 15th September 1956. It was transferred to the Southern Region at Dover Marine in December 1959, moving west to Exmouth Junction via Ashford and Tonbridge. A move to the Western Region's Templecombe (S&D) shed followed in late summer 1964. A move to Bristol Barrow Road, in June 1965, was followed by a final transfer to Bath Green Park (S&D) a month later. It was withdrawn on 18th November 1965 and demolished by Buttigiegs, Newport, the following January.
*A. R. Carpenter*

No 80060 stands on shed at Kentish Town whilst allocated to Bedford depot. New to Bedford on 19th March 1953 it was to remain there until January 1955 when it was moved to Bury. A year later it would be at Newton Heath where it would stay until 29th February when it was reallocated to the Scottish Region at Stirling. Moves via St Margarets and Greenock Ladyburn would see its arrival at Polmadie on 13th December 1965 from where it would be withdrawn on 7th February 1966. Kentish Town depot was opened by the Midland Railway on 8th September 1867 and closed by BR in April 1963. Adjacent to the depot was Read Brothers Bottling Stores – before the late 1860s, Burton brewers supplied London by sending their beer via the Midland Railway's competitors. However, when the Midland planned its main line to London in the early 1860s, Bass agreed to send all their beer with the company for a fixed price. In return the Midland would provide 'Ale Stores and Offices sufficient for the business' at St Pancras. The railway built a dedicated warehouse adjacent to the Regent's Canal that was connected to St Pancras's northern goods yard. This held 120,000 barrels and employed 120 men. In 1874 it sent 292,300 barrels of beer to London, 36% of its total output! By the mid-1960s, beer traffic to St Pancras had ceased.

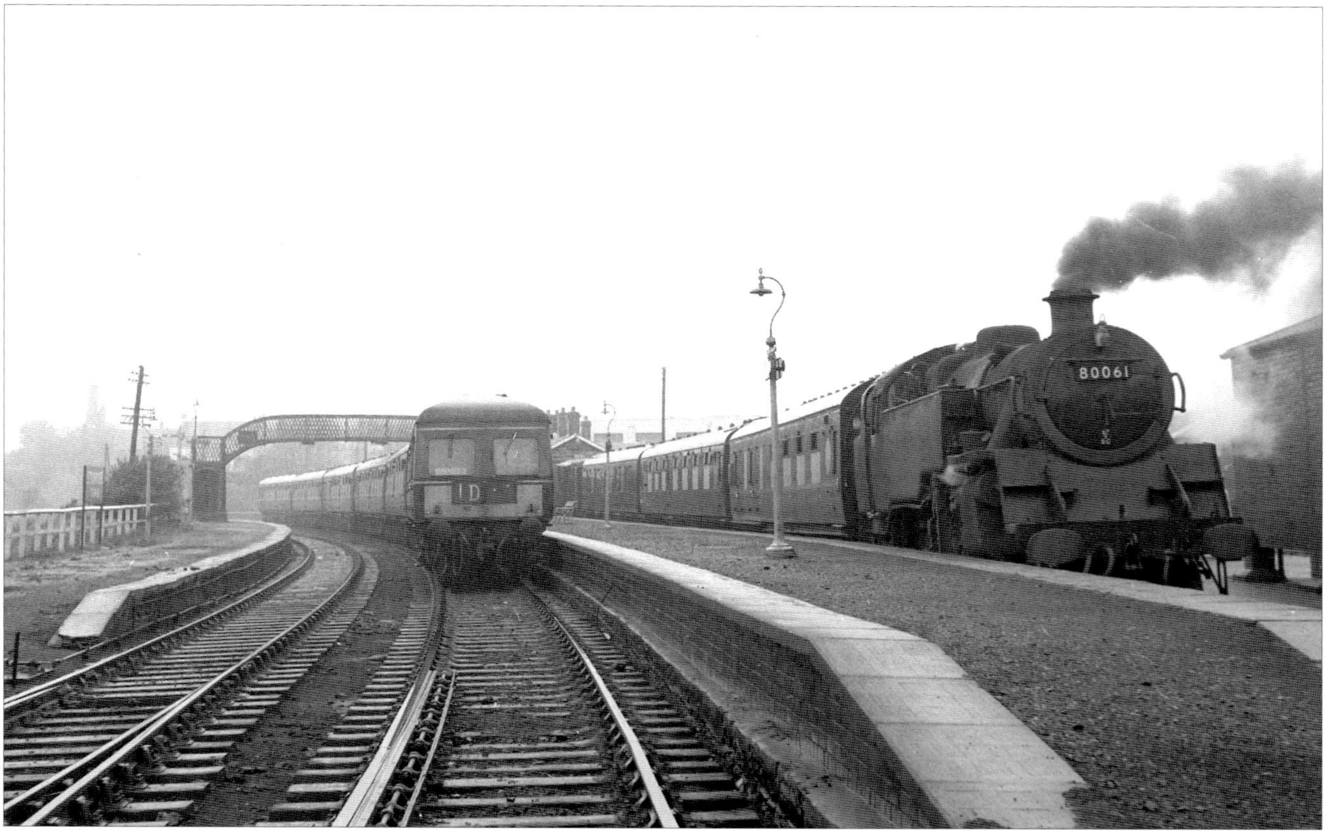
80061
1D

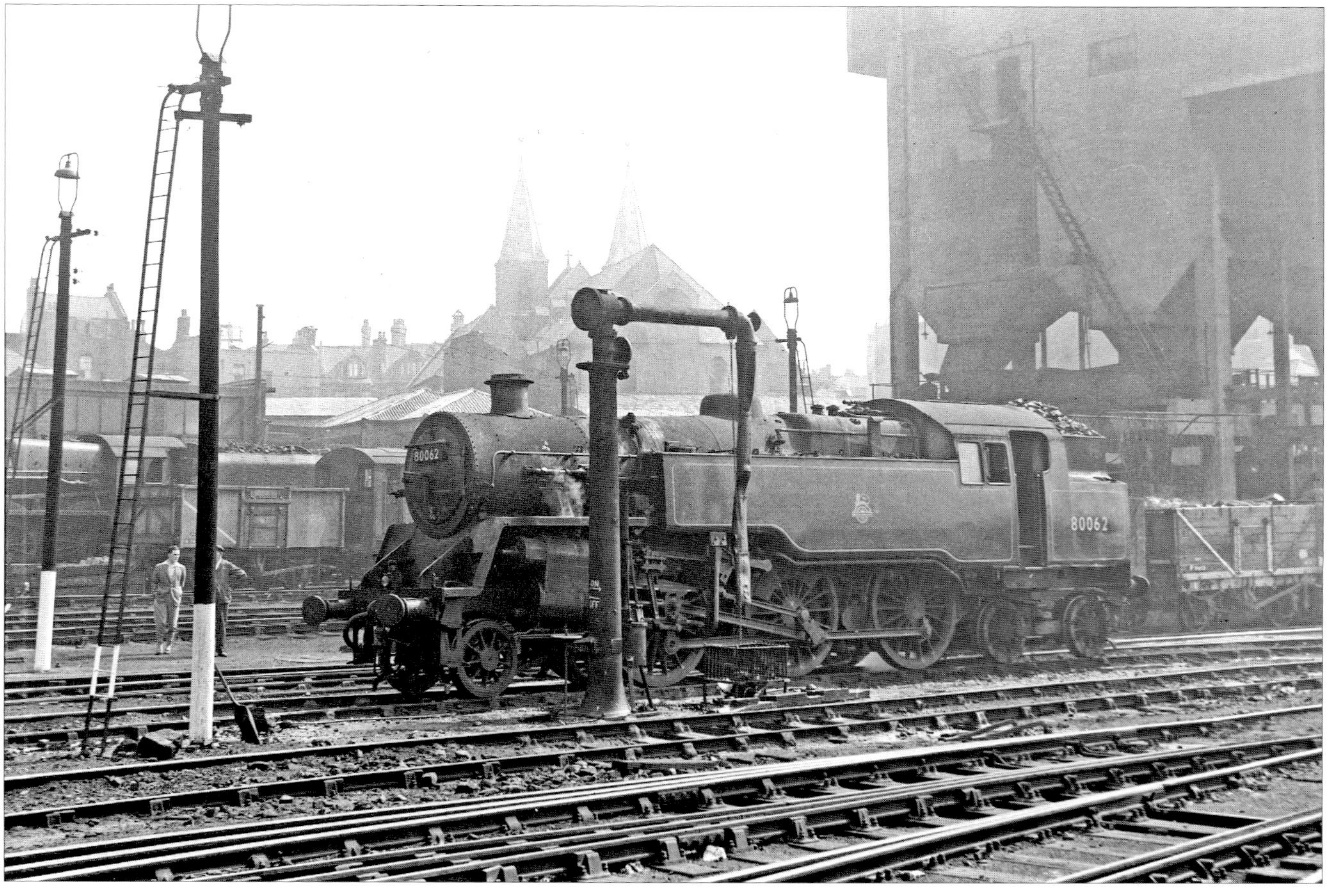
80062
80062

# The Class of 1953

Top left. At Stranraer Town on 22nd May 1965 No 80061, transferred the previous August from Stirling, and one of the 19 of the class transferred to Scotland from the north west of England in early 1960, awaits departure for Dumfries on the 3.50pm 'all stations' on the former Portpatrick and Wigtownshire line which would see its final trains three weeks later. At the main platform is a Swindon Inter City diesel multiple unit forming the 4.25pm to Glasgow St Enoch. Stranraer Town, known simply as Stranraer until 2nd March 1953 and which closed on 7th March 1966, had only two arrivals and three departures, two of which are seen in this photograph, each day in early 1965. The station had opened with the rest of the Portpatrick Railway west from Castle Douglas on 12th March 1861 and from 28th August 1862 also offered services across the Rhins of Galloway to Portpatrick, initially considered to the the port for Ireland but soon supplanted by Stranraer Harbour which station opened on 1st October 1862 and remains open with services north to Girvan and Ayr. Passenger services between Stranraer and Portpatrick were withdrawn on 6th February 1950, with goods traffic remaining to the creamery at Colfin until 1st April 1959. *J. L. Stevenson*

Bottom left. No 80062 was new to traffic on 29th April 1953 and is seen on shed at Kentish Town four months later on 9th August. The sheer bulk of the concrete coaling stage almost overpowers the locomotive. No 80062 left London, bound for Chester, in September 1956. It moved to Birkenhead Mollington Street early in 1958 before a move to the Scottish Region. The locomotive arrived at Greenock Ladyburn in the summer of 1960 with a final move to Stirling in August 1964 from where it was withdrawn two months later on 12th October. *A. R. Carpenter*

Below. The scene is Comrie on 4th July 1964 with No 80063 waiting to depart with the 6.45pm service for Gleneagles – this was the last day of passenger services, freight services had already ceased on the 15th June. The Crieff & Comrie opened in 1893, connecting Comrie to the railway network at Crieff. The tourism potential of Loch Earn was an important factor, and the route was later extended westward to Lochearnhead. However the line was never successful, and declined in the 20th century, particularly due to cheap and frequent bus competition. Four-wheel rail buses were introduced in 1958 to reduce operating costs, but the decline continued and the line closed. No 80063 entered traffic from Brighton Works on 13th May 1953, being allocated to Saltley. It arrived in Scotland in spring 1960, being allocated to Stirling – a final move to Corkerhill came in the summer of 1964 with withdrawal following on 24th August 1966. *W. A. C. Smith (WS7646)*

43

# The Class of 1953

Top left. No 80064 is seen on 27th March 1955 between duties at Bletchley depot. Allocated new to Watford depot on 9th June 1953, it moved south to Dover Marine in December 1959 with stays at Ashford and Tonbridge before moving west to Exmouth Junction in June 1962. A final reallocation to Bristol Barrow Road came in May/June 1965 from where it was withdrawn on 25th August the same year. The locomotive was sold to scrap merchants Woodham Brothers who operated out of Barry Docks. It was one of the class to be sold into preservation and arrived at the Dart Valley Railway (Buckfastleigh) in February 1973. The locomotive was restored to steam in February 1981. It can now be found on the West Somerset Railway awaiting overhaul. The LNWR opened its Bletchley engine shed in 1851, the original building was blown down during a storm in 1872. A replacement was opened in 1873 and following re-roofing by BR it closed on 15th July 1965 and was demolished. *E. Sawford (ES2221)*

Bottom left. New from Brighton Works on 25th July 1953 No 80065 was initially allocated to Watford, remaining there until transferred to Stewarts Lane in December 1959. Following allocations to Ashford and Tonbridge it arrived at Eastleigh in June 1962 where it would remain until withdrawn on 4th September 1966. It is seen here at Hammersmith & Chiswick having arrived with the Railway Correspondence & Travel Society's railtour down the branch from South Acton Junction. Passenger services on the branch had ceased on 1st January 1917 – freight would continue until 3rd May 1965. One of the more intrepid enthusiasts appears to be taking a dive off of a pile of tyres. *F. W. Goudie (703)*

Below. Another of Watford's allocation of Standard Fours is seen at Bletchley depot on 15th July 1954, the locomotive is seen at the depot's coaling stage that was surmounted by a water tank. New to traffic exactly a year earlier, the locomotive was reallocated to the Southern Region's Stewarts Lane depot, following the same route taken by No 80065. It would be withdrawn on 13th June 1965 and probably towed along with No 80065 to Birds scrapyard at Morriston in South Wales. *E. Sawford (ES1493)*

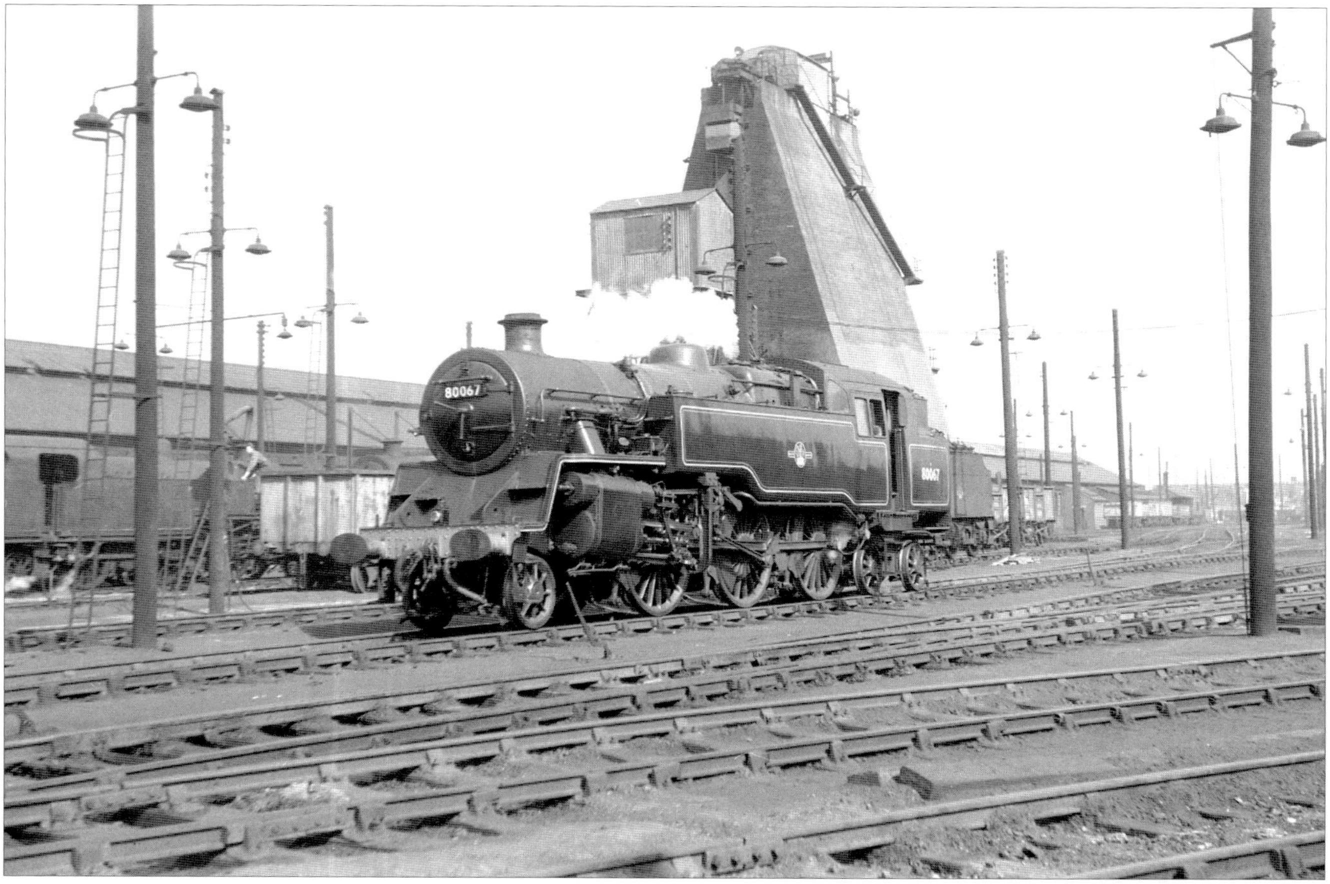
80067

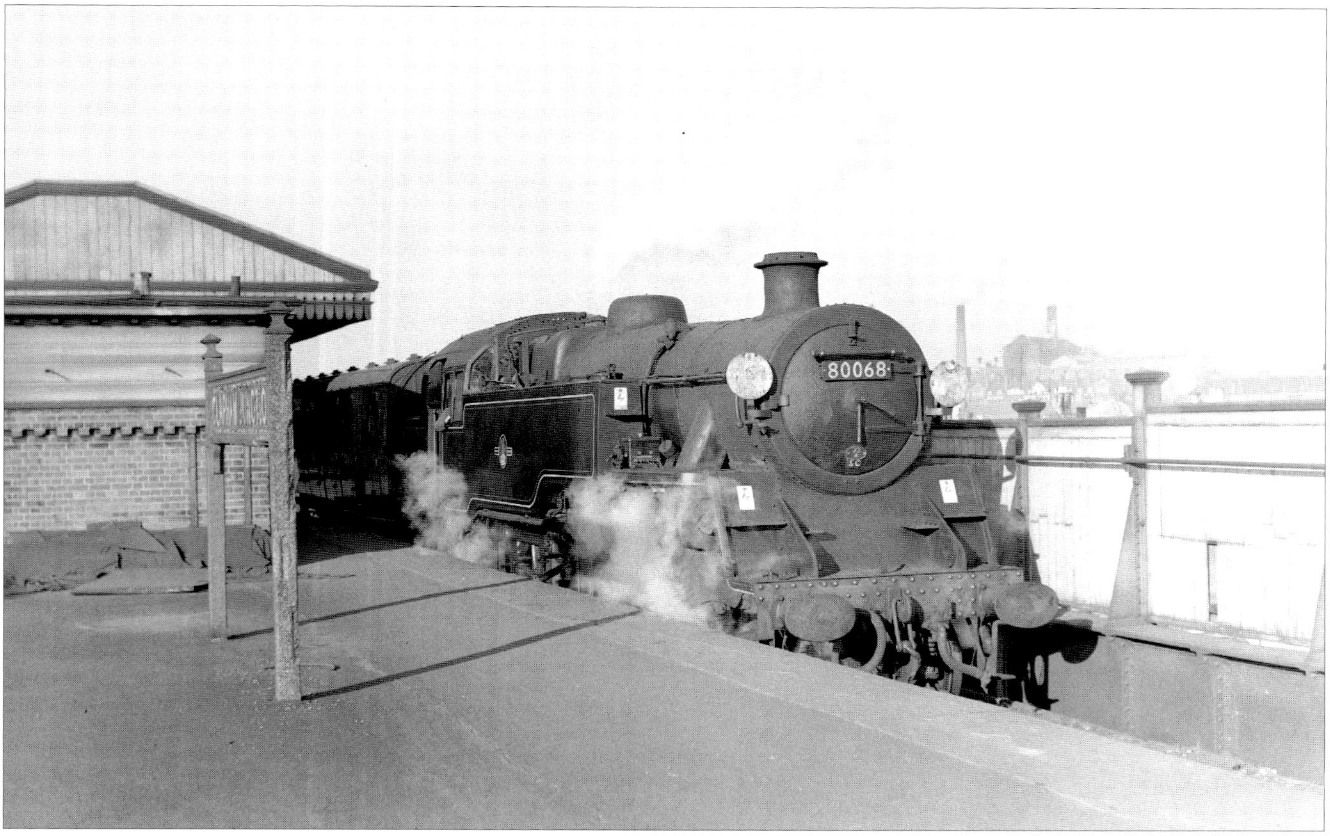
80068

# The Class of 1953

Top left. With the concrete coaling tower dominating the background, No 80067 is seen on shed at Willesden on 23rd April 1959. The LNWR opened the original depot in 1873; facilities consisted of a 12-road dead end shed with turntable, water tank and coaling stage. It was enlarged in 1898, but the roof was removed in 1939 due to its poor condition. A second shed had been constructed in 1929, remaining open until closed by BR or 27th August 1965. No 80067 arrived at Watford depot on 4th August 1953 where it remained until December 1959 when it was transferred to Stewarts Lane. It moved west in late summer 1965 when it was transferred to the former S&D depot at Templecombe. It was withdrawn on 3rd June 1965 and sold to Woodham Brothers at Barry – unlike numerous other members of the class it did not survive to be preserved. *A. Swain (E84/2-3)*

Bottom left. No 80068 stands at Clapham Junction on 17th March 1961. The locomotive will be departing with the 8.15am to Olympia (Addison Road). Unusually this service was not to be seen in the published timetable as it was for Post Office clerical workers – it would be the last 'branch line' in London to be operated by steam. New to traffic on 21st April 1953, No 80068 had arrived at Stewarts Lane, via Watford, in December 1959. It would move to Brighton in September 1963, before moving via Redhill to Feltham where it was allocated in May 1965. It was withdrawn on 2nd October 1966 and sold to Cashmore's, Newport, for recycling. *J. Harrold (H2058)*

Below. Allocated new to Plaistow on delivery from Brighton Works on 30th September 1953, No 80069 saw service at Tilbury and Swansea East Dock before being reallocated to Nine Elms on 9th August 1964. It is seen at Norbiton during March 1965 at the head of a Wimbledon-Feltham freight train consisting of milk tanks. As the trip working will include a reversal at Twickenham, the consist has a brake van at both ends. Withdrawn on 23rd January 1966, it was scrapped by Cox & Danks at Park Royal in April 1966 – one of six to be recycled at the yard. *L. Fullwood (FS25/2)*

Above. No 80071 is seen at the head of a LT&SR service whilst allocated to Tilbury depot where it had been allocated to in January 1954. It had been new to Plaistow on 21st October 1953. A transfer to Stratford took place in July 1962 where it stayed a matter of weeks before moving north to March depot in September. This was followed by a move to the North Eastern Region followed in December 1962 when it was allocated to Ardsley. A final transfer to Carstairs took place 10 months later; following withdrawal on 30th July 1964 it was sold to Motherwell Machinery & Scrap, Wishaw, for recycling – it was broken up the following February. *R. E. Vincent (REV162-3)*

Top right. Delivered from Brighton Works, No 80074 was allocated new to Plaistow on 27th November 1953. It was reallocated to Tilbury just before Christmas 1956. It is seen here a few months later, on 16th March 1957, at Rainham, Essex. The station had been opened by the LT&SR in 1854 – it was on the original route from London to Tilbury, which was extended to Southend-on-Sea in 1856 and Shoeburyness in 1884. In 1888 a second, more direct, route to Southend was completed several miles to the north. No 80074 was transferred to Stratford in July 1962 and followed the route No 80071 took to Scotland including its demise in Wishaw. *J. Backler (JB000448)*

Bottom right. The 11.40am service from Fenchurch Street is seen arriving at Tilbury behind No 80075 on 17th March 1956. New to Plaistow on 10th December 1953 it took the by now familiar route to Carstairs, being scrapped in Wishaw during February 1965. Tilbury depot's coaling plant can be seen above the second carriage, note that the inner walls of the turntable pit have been painted white. The depot closed on 18th June 1962 and was demolished soon after. *L. R. Freeman (F2178)*

Above. No 80076 is seen at Barking on 15th September 1956 with the 1.14pm Tilbury-Fenchurch Street service. The station was opened on 13th April 1854 by the LT&SR as one of the first stations on the route. At this time Barking was a small village and the original station was a two platformed affair. It was opened as part of the LT&SR's new line that left the Eastern Counties Railway's (ECR) main line at a new junction at Forest Gate. It was rebuilt in 1908 and again in 1959. At the time of writing, significant redevelopment of the station is on-going. No 80076 emerged from Brighton Works and was allocated to Plaistow on 23rd December 1953. It left the LT&SR in July 1962 when it was reallocated from Tilbury to Stratford. A move to March depot occurred that September and a further move in December saw it reallocated to the North Eastern Region, being allocated to Ardsley. A final move to Scotland came in October 1963 when it was reallocated to Dumfries from where it was withdrawn on 30th July 1964. *L. R. Freeman (F2384)*

Top right. The first Class 2MTs left Crewe Works in the summer of 1953. No 84000 was allocated new to Crewe North on 4th July. This image is dated 8th August 1953 by which time the locomotive was allocated to Bolton's Plodder Lane depot – the locomotive has yet to receive its 10D shed plate so is probably seen shortly after transfer. The LNWR's Plodder Lane shed was opened on 1st April 1875. The original building was closed in 1944 and demolished in 1950 and replaced with a turntable. The LNWR added a second shed in 1891. This was closed on 10th October 1954 and later demolished. No 84000 was relocated to Wrexham Rhosddu in March 1954 with subsequent moves to Birkenhead Mollington Street and Warrington Dallam. A change of regions occurred in spring 1963 with a move to the Western Region's Oswestry depot. It moved back to the London Midland Region early in 1965 with a move to Wrexham's Croes Newydd depot from where it was withdrawn on 30th October 1965. *A. G. Ellis*

Bottom right. The fitting of push-pull apparatus to the class is seen to good effect as No 84001 and its two-coach train travel from Amlwch towards Bangor late on in its career whilst allocated to Llandudno Junction. It had arrived there in November 1962 and was withdrawn on 31st October 1964 being sold to Hughes Bolckows at North Blyth for demolition. It had arrived at Crewe North on 6th July 1953 from the nearby works. During its career of just over 11 years it spent time at Plodder Lane, Wrexham Rhosddu, Chester Northgate, Birkenhead Mollington Street and Warrington Dallam. Opened by the Chester & Holyhead Railway on 1st May 1848 Bangor station was lies between Bangor Tunnel to the east of the station, and Belmont Tunnel to the west, the station was progressively expanded into a junction station as a number of branch lines were opened: From Menai Bridge to Caernarvon (Bangor & Carnarvon Railway) (1848); from Gaerwen to Amlwch (Anglesey Central Railway) (1866); from Holland Arms to Red Wharf Bay and Benllech (Red Wharf Bay branch line) (1909); to Bethesda (Bethesda Branch) (1884). With the closure of the branch lines in the 1960s and 1970s, the station was reduced to just two operational platforms. *E. N. Kneale (KN72/2)*

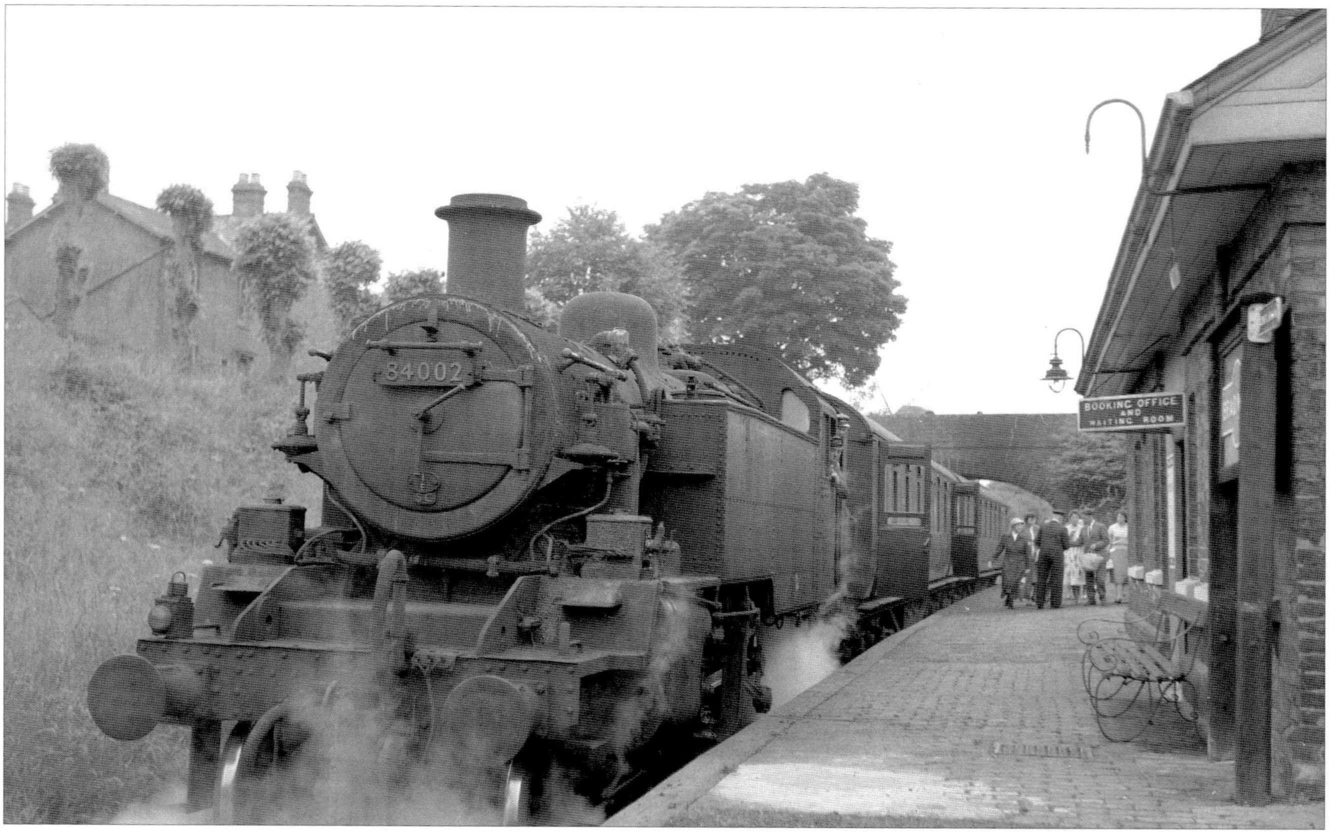
84002
BOOKING OFFICE AND WAITING ROOM

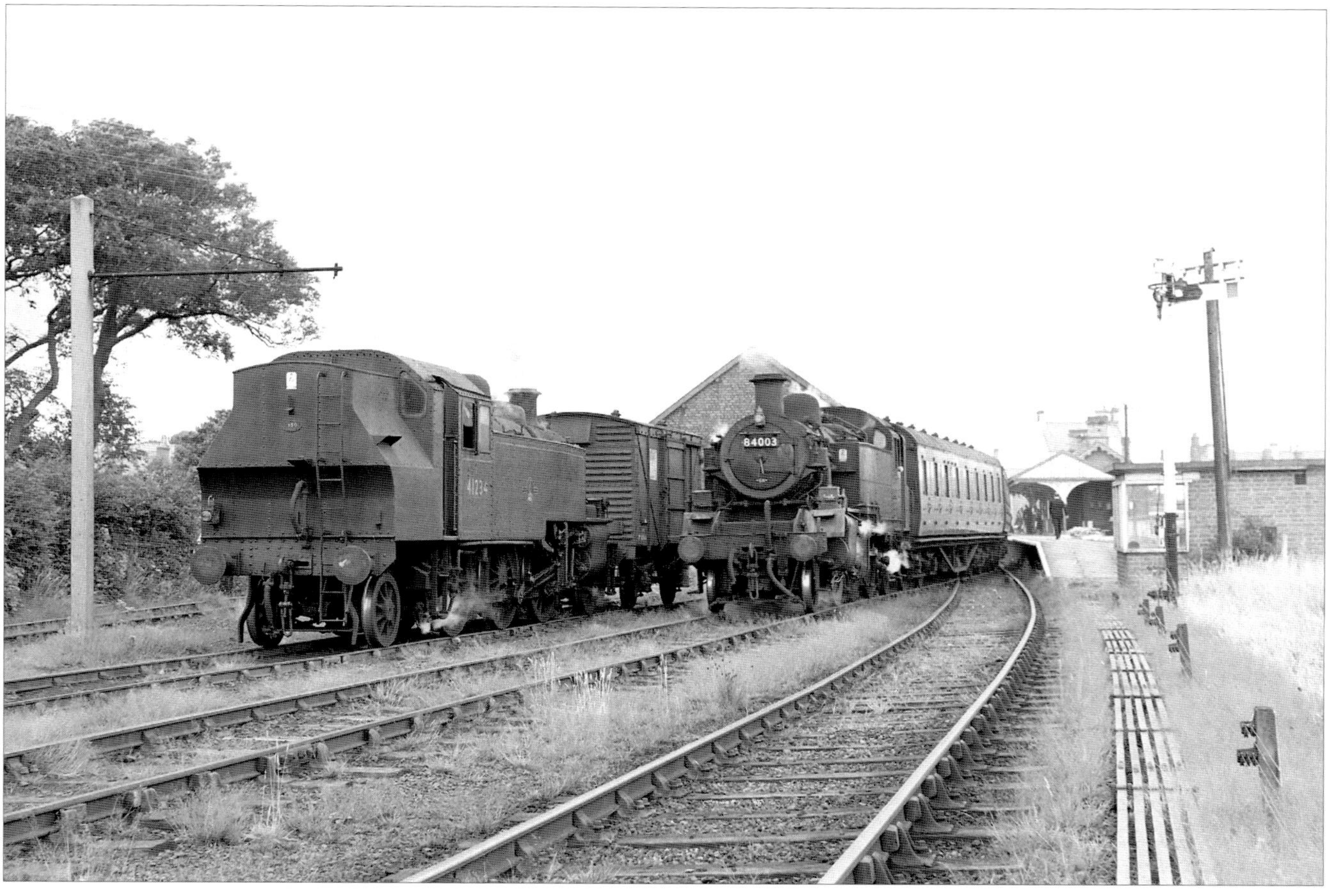
84003

# The Class of 1953

Top left. The former LNWR's Bradwell station is seen with No 84002 awaiting departure for Newport Pagnell on 27th June 1956 . It served both Bradwell and the new village of New Bradwell in Buckinghamshire. The station, which consisted of a brick-built station building, and single platform, opened to traffic on 2nd September 1867. The last passenger train ran on 5th September 1964 but freight trains continued to pass through until 22nd May 1967. The station building was demolished although the platform remains intact. The track bed through the station has been converted into a shared path (footpath/cycle way), forming part of the Milton Keynes redway system. No 84002 was new to Plodder Lane on 7th August 1953 with stays at Wrexham Rhosddu and Chester Northgate before arriving at Bletchley during the summer of 1956. It was withdrawn from the latter depot on 17th April 1965 and consigned to Buttigiegs, Newport, for scrapping. *J. Harrold (H806)*

Bottom left. Cousins meet up at Amlwch – the BR Standard Class 2MTs were based on Ivatt's design for the LMS and both are seen here. No 84003 was new to traffic at Bolton's Plodder Lane depot on 10th August 1953; moving onto Wrexham Rhosddu the following March and then Birkenhead Mollington Street in the summer of 1956 A reallocation to Rhyl occurred in late summer 1961, it then spent two periods allocated to Llandudno Junction – February 1963 to mid March 1964 and early June to 2nd October 1965 when it was withdrawn. No 41234 was new to traffic from Crewe Works in August 1949 being withdrawn on 19th November 1966 – it had spent a short period at Llandudno Junction – w/e 19th June 1965 to w/e 7th August 1965, conveniently giving a time frame to the image. Amlwch was the original terminus of the Anglesey Central Railway line from Gaerwen. A light railway extension was later added for freight purposes. All stations on the Amlwch line closed to passengers on 7th December 1964 following the Beeching Report. The line was kept open for rail traffic from the Associated Octel works with tankers of brominated products – this included Dibromeothane, a key component of leaded petrol. The freight traffic continued until 1993 when road tankers took over, with the works closed in March 2004. *E. N. Kneale (KN851*

Below. No 84004 was new to Plodder Lane depot on 19th August 1953 and is seen here at Wrexham Central whilst allocated to the town's Rhosddu depot. It had arrived in the town during late summer 1954 staying until a move to Bletchley in the summer of 1956. A move to the Western Region at Oswestry came in April 1963, returning to the LMR at Wrexham's Croes Newydd depot from where it was withdrawn on 30th October 1965. Rhosddu closed to steam on 4th January 1960 and subsequently demolished. The original Central station was opened by the Wrexham, Mold & Connah's Quay Railway on 1st November 1887. The WM&CQ was formally taken over by the Great Central Railway on 1st January 1905. The station closed to all traffic on 23rd November 1998 and was demolished, todays station is located further to the west. *N. Stead collection (NS207848)*

Top left. No 84005 is seen departing Northampton Castle station on 28th February 1962. At one time there were three stations in the town: Bridge Street (LNWR), St John's Street (MR) and Castle; the latter now survives as the town's only station situated on a loop off the West Coast main line. Opened by the LNWR on 16th February 1859 it took its name as it was built on the site of the former castle, it lost the Castle suffix on 18th April 1966. No 84005 was new to Bedford on 24th August 1953 and was to spend three periods allocated there – new to June 1961, December 1961 to March 1962 and a matter of days in February 1962. It spent its 12 years of service shuttling round various LMR depots, with multiple stints at Neasden, Wellingborough and Leicester Midland. It was withdrawn from the latter depot on 30th October 1965. *Alec Swain (L51/1-1)*

Bottom left. Wellingborough station was built by the MR in 1857, on its extension from Leicester to Bedford and Hitchin. At the time, the station was known as Wellingborough Midland Road to distinguish it from one built by the LNWR in 1866, at Wellingborough London Road for the Northampton & Peterborough Railway, which closed to passengers on 4th May 1964. A curve linked the two stations from west to north. No 84006 was new to Burton depot on 26th August 1953, relocating to Wellingborough in January 1959 – carrying shed plate 15A the image is dated to this period of time up to December 1961 when it was reallocated. This was followed by transfers to Neasden (twice), Bedford, Annesley, with a return to Wellingborough between July and September 1964. Withdrawal came at Leicester Midland depot on 30th October 1965. The writing is on the wall for steam as in the adjacent platform is DMU Class 127 Motor Brake Second No 51598. *J. A. C. Kirke*

Above. No 84007 was new to Burton on 31st August 1953 and is seen here at Higham Ferrers on 6th June 1959, having been transferred to Wellingborough depot in early February that year. Higham Ferrers was the terminus of an MR branch from Wellingborough, being opened on 1st May 1894. Initially named Higham Ferrers, it was renamed Higham Ferrers & Irthlingborough on 1st July 1902, but reverted to its original name on 1st October 1910. It closed to passenger services on 15th June 1959, but remained open for a further 10 years as a private siding. No 84007 would be transferred to Annesley in mid-March 1962 with a return to Wellingborough in early December 1963. It was withdrawn on 18th January 1964 and sold to Albert Looms, Spondon, for demolition. Today the company is the largest car dismantler and vehicle recycler in Derbyshire and Nottinghamshire.

Above. No 84008 was new to Burton depot on 31st August 1953, remaining there until early February 1959 when it was reallocated to Wellingborough, dating the image to this period of time. The original station, called simply Tutbury, was opened on 11th September 1848 by the North Staffordshire Railway. Nestlé have a historical presence in the village of Hatton due to the surrounding farmland, which supported a strong dairy farming industry. Until the late 1970s the factory had its own private siding, which gave access to milk trains from the station. The original Tutbury station closed to passengers on 7th November 1966, general freight traffic had ceased on 6th July 1964. The present station was opened on 3rd April 1989 and serves the villages of Tutbury, in Staffordshire, and Hatton in Derbyshire. No 84008 was subsequently allocated to Neasden, from early June 1961; Kentish Town from early July 1962 then rotated between Leicester Midland and Wellingborough. Destined for South Wales for scrap, it was withdrawn from Leicester Midland depot on 30th October 1965. *R. E. Vincent (REV466)*

Top right. No 84010 was allocated new to Low Moor on 8th September 1953 where it would stay for a year before moving to Oldham's Lees depot. It moved to Rose Grove in May 1956 and probably received the second BR emblem during a Light Intermediate overhaul at Crewe Works late in 1957. It was reallocated to Fleetwood depot in early December 1959. This was one of the class destined to be transferred to the Isle of Wight; it was reallocated to Eastleigh in early December 1965 but would not make the move south as plans changed and it was returned to Fleetwood for immediate withdrawal on 11th December 1965. *R. Bruce*

Bottom right. Resting between duties on Aintree shed is No 84012 during the time it was allocated to Bank Hall depot. It was new to Low Moor on 17th September 1953. Reallocations to Lees (Oldham) and Rose Grove followed with it arriving at Bank Hall in late summer 1955. It was to stay there until moving to Lower Darwin in spring 1957. Moves to Fleetwood followed in mid-March 1961 and to Southport in February 1963. It would be withdrawn, as the first of the class to be condemned, from the latter depot on 12th October 1963, and broken up at Crewe the following month after 10 years and 25 days in service. Aintree depot was opened by the Lancashire & Yorkshire Railway in 1886. The LMS reroofed the shed and reorganised the yard in 1937 with new ash and coaling plants, a resited turntable. Closed on 12th June 1967 it stood trackless and derelict for another 30 years, finally being demolished in February 1996. *N. Stead collection (NS207833)*

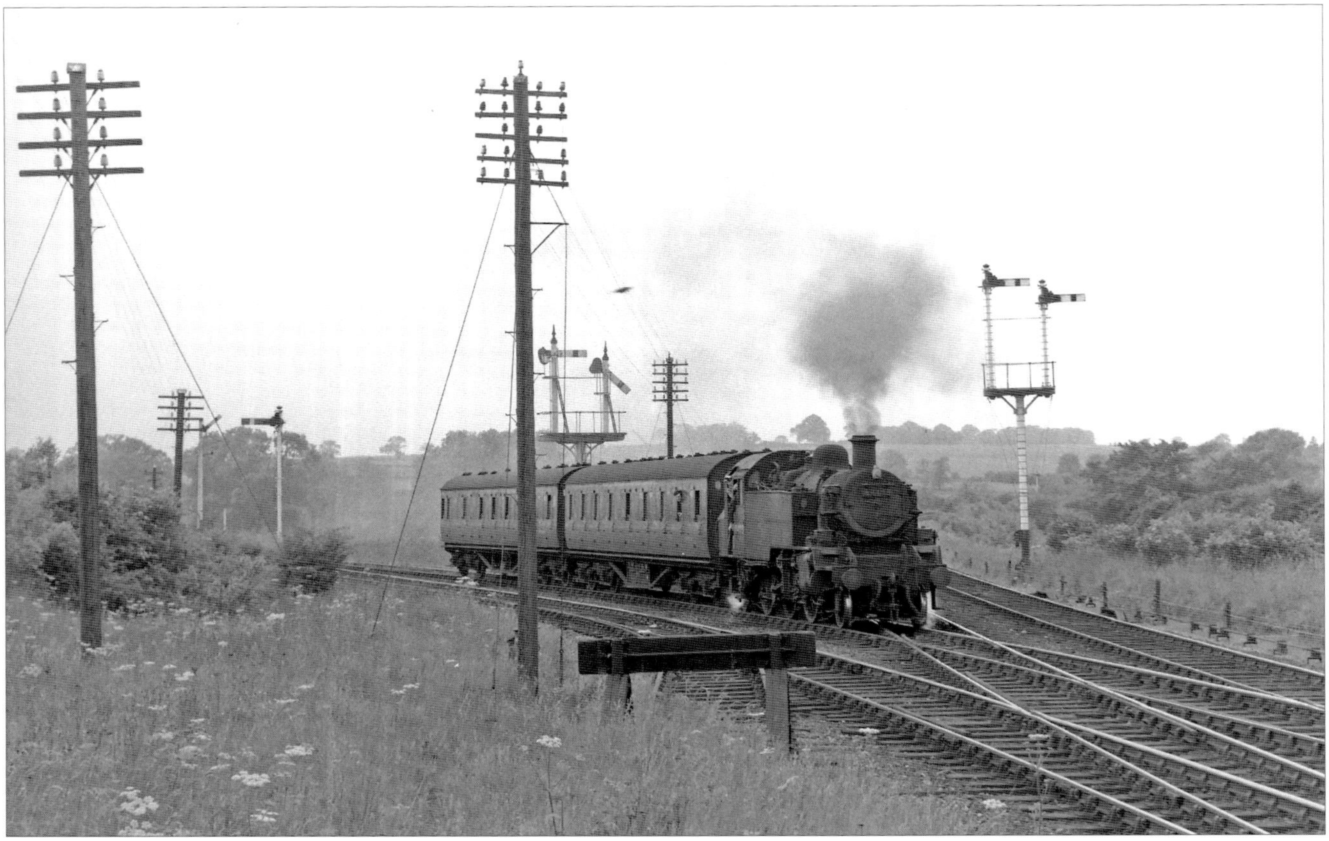

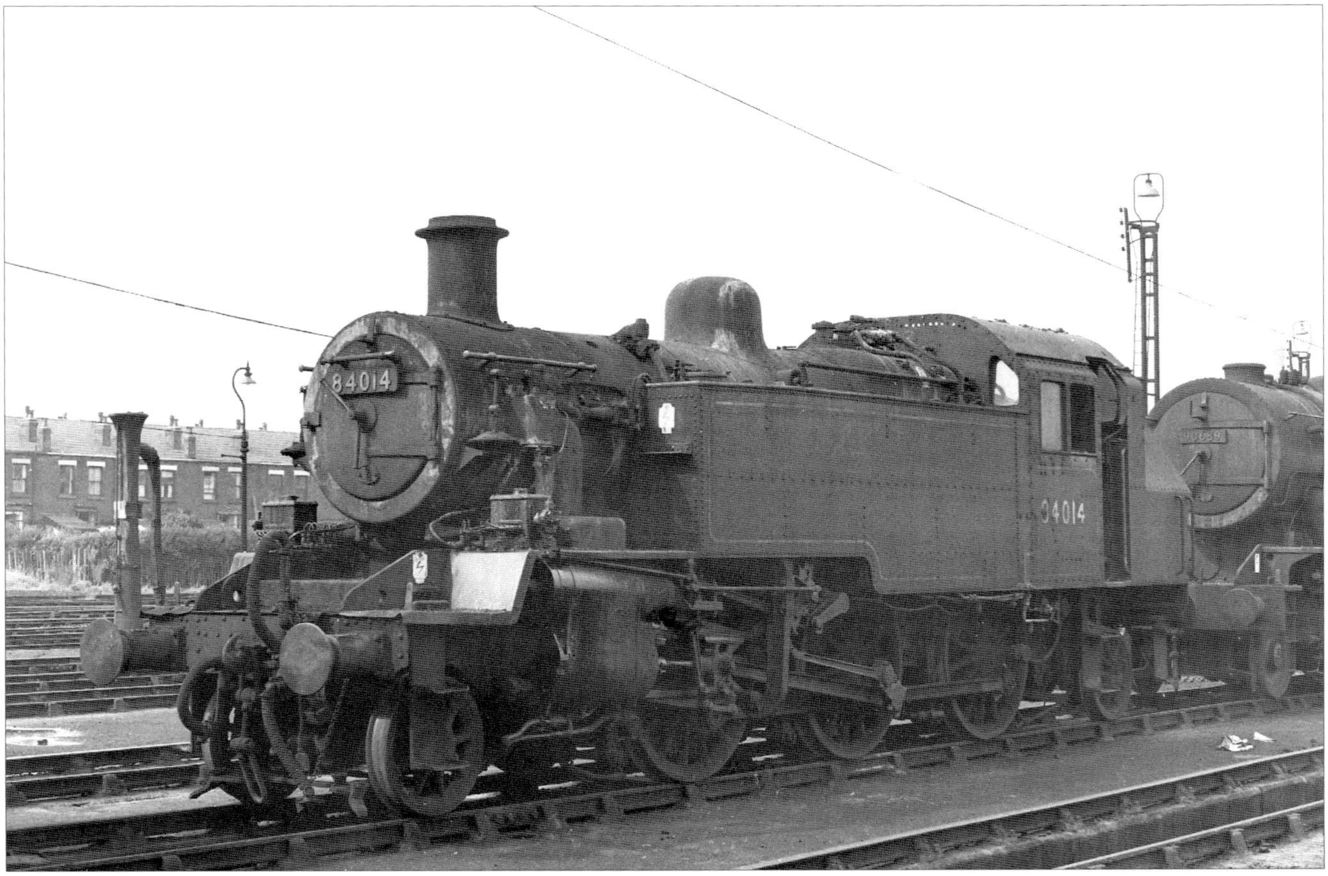
84014

# The Class of 1953

Top left. The 12th June 1965 was an overcast day as we see No 84013 at Luffenham junction with the 2.50pm Seaton-Stamford service. The locomotive was new to Low Moor depot on 21st September 1953. In January 1955 transfer to Bank Hall was followed by moves to Lees (Oldham), Bolton, Stockport Edgeley and to Leicester Midland in mid-May 1965 where it was allocated at the time of the photo. A return to Stockport Edgeley took place in early July. A paper transfer took place for reallocation to Eastleigh, as one of those destined for the Isle of Wight, this did not take place and the locomotive was withdrawn from Stockport Edgeley on 11th December 1965. Seaton was originally a minor intermediate station on the LNWR's single track Rugby-Stamford line, which opened throughout in 1851 – at the time the station was named Seaton (Uppingham). On 1st October 1894 the branch line to Uppingham was opened, with Seaton losing the Uppingham in its title. Seaton became a junction in 1873 when the LNWR double tracked the line from Rugby to Seaton and opened a new double track line thence to Wansford. The Rugby to Peterborough was then operated as the main line and Seaton to Stamford as a branch line. General goods traffic ceased on 4th May 1964 with passenger traffic being withdrawn on 6th June 1966, at which point the line closed completely. *M. Mitchell (MM2872)*

Bottom left. The original Bolton depot was opened by the L&YR in 1874, it was enlarged in 1888. It was reroofed by the LMS in 1946 and became one of the last steam sheds to be closed, in this case on 1st July 1968. No 84014 is seen there on 13th September 1964 in the company of Austerity 2-8-0 No 90689. The 2MT was new to Low Moor on 22nd September 1953, and arrived at Bolton, via Bank Hall, Southport, Lees (Oldham) in mid-February 1965. It remained at the depot until mid-December 1964 when it was allocated to Stockport Edgeley. In early November 1965 it was reallocated to Eastleigh as one of the LMR batch for the Isle of Wight – it was the only member of the LMR locomotives destined for the island to reach the Southern Region. Although nominally returned to Stockport it was withdrawn on 11th December 1965, destined for scrapping at Cashmore's Newport scrapyard. *K. Nuttall (KNL271)*

Below. No 84016 is seen entering Thornton-Cleveleys station on 21st August 1964 with the 7.25am Blackpool North-Fleetwood service. The original station was opened in April 1865 by the Preston & Wyre Joint Railway, being named Cleveleys. It was to the south of Station Road in Thornton, near an older halt called Ramper Road. On 1st April 1905 the station was renamed Thornton for Cleveleys. This station closed in 1927 when the new station (the first to be built by the LMS) opened to the north of the level crossing. In February 1953, the station was renamed again, this time to Thornton-Cleveleys. Rationalised in the 1950s and 1960s, and affected by the ending of the ferry from Fleetwood to the Isle of Man, the station eventually closed on 1st June 1970, when the Fleetwood line was closed to passengers. Freight continued on the line to nearby Burn Naze until 1999. New to Bury depot on 7th October 1953 No 84016 arrived at Fleetwood shed in early summer 1954. In the early summer of 1957 it was reallocated to Lancaster Green Ayre before a return to Fleetwood in mid-November 1958 where it stayed, apart from a month at Stockport Edgeley, until its nominal transfer to the Southern Region in November 1965. In reality it was dumped at Lostock Hall labelled 'Dead engine to Fratton, Portsmouth', withdrawal took place on 11th December 1965. *L. Fullwood (FM47-2)*

No 84018 was delivered from the works on 23rd October 1953 and is seen here in the shed yard on 1st November 1953 running in before despatch to Bury depot. In early summer 1954 it was transferred to Fleetwood where it was to remain for the rest of its career. Condemned on 24th April 1965, it was sent for scrapping to Hughes Bolckows yard at North Blyth. *A. W. Battson*

In 1954, Willesden in North London was the site of a joint exhibition by British Rail, London Underground and various manufacturers for the International Railway Congress. The joint exhibition included stock from London Underground, various pieces of permanent way & signalling equipment, including some of the UK's newest tech. At the time containerised transport was in its infancy having been pioneered by UK railways with wooden containers that were craned between rail wagons and road transport, and several examples were on display. The displays included Wickham railcars for maintenance crews. The locomotives on show ranged from prototype gas turbine and BR Class 8P No 71000 Duke of Gloucester, the latter was so new that no photos were available for the brochure, instead a line drawing was added. One of the exhibits was No 84019 that was ex works at Crewe on 31st October 1953. New to Bury it arrived at Fleetwood in early summer 1954, moving on to Lees (Oldham) and Bolton where it arrived in spring 1958. Another of the non-arrivals on the Southern Region it was condemned on 11th November 1965 it was to be broken up by Arnott Young's cutters in Parkgate the following March. N. Stead collection (NS207843)

# The Class of 1954

Above. No 80077 is seen arriving at the unstaffed Barleith station that was near Hurlford, to the east of Kilmarnock, East Ayrshire, Scotland. The station was built by the Glasgow & South Western Railway on its Darvel branch. The line between Darvel and Ryeland on the route to Strathaven had closed on 11th September 1939. Barleith station was opened some time before 1904, it was renamed Barleith Halt in 1944, but reverted back to Barleith in January 1954. The station closed permanently to passengers on 6th April 1964, the final trains ran two days earlier and we see No 80077 arriving with its single-coach train with a couple of enthusiasts hanging out of the open door droplights. The locomotive was completed at Brighton Works on 12th January 1954 and sent to the LT&SR's Plaistow depot. It left the Eastern Region in December when it was reallocated to the NER's Ardsley depot. A move to Scotland followed in October 1963 with its arrival at Ardrossan. Declared surplus to requirements after the closure of the Darvel line it was reallocated to Corkerhill from where it was withdrawn on 12th October 1964. *W. A. C. Smith (7367)*

Top right. Seen amongst the general untidiness of a steam shed, No 80078 is seen at Stratford depot on 14th March 1958. It had been allocated new to the LT&SR's Plaistow depot on 2nd February 1954, arriving at Stratford depot via Tilbury in June 1962. It was reallocated to the Western Region's Shrewsbury depot on 15th July 1962. A move to Wrexham in February 1963 saw its withdrawal from the town's Croes Newydd depot on 24th July 1965; it was one of the fortunate examples to be sent to Woodham's Barry scrapyard to be preserved.

Bottom right. No 80080 was new to traffic on 18th March 1954, being allocated new to Plaistow depot. Reallocated to Tilbury in May 1954 it made its way to Croes Newydd along with Nos 80078/79 in July 1962. It is seen outside the erecting shop at Crewe Works; it had arrived for a Heavy Intermediate overhaul in early February 1964 that was completed five weeks later, by this time Croes Newydd was LMR property and the locomotive sports the depot's 6C shed plate. Withdrawn on 24th July 1965, it may well have accompanied No 80078 to South Wales as it has also survived into preservation. *E. N. Kneale (KN835)*

Top left. On a snowy 27th February 1955 No 80081 is seen on shed at Bletchley where the locomotive had been allocated on release from Brighton Works on 31st March 1954. Transferred to Willesden in early November 1959, the locomotive was sent south of the Thames in December 1959 when it arrived at Stewarts Lane. In July 1963 it went west to Weymouth Radipole depot – by this date the former LSWR shed was no more and all locomotives were concentrated at the former GWR depot that was to close with the end of steam on the Southern on 7th July 1967. No 80081 was not to stay long, being reallocated to Bournemouth the following month. It was withdrawn on 8th June 1965, and disposed of by Bird's cutters at Morriston. *E. Sawford (2069)*

Bottom left. A local service from Southampton Central to Bournemouth is seen departing 'Soton' behind No 80082 on 20th March 1966. The locomotive was new to Bletchley on 15th April 1954, moving south to Bricklayers Arms in December 1959. It had arrived at Eastleigh in June 1962, from where it would be withdrawn on 4th September 1966. Southampton Central station was opened as Southampton West in 1895, to replace the smaller nearby West End station (originally named Blechynden when it opened in 1847). The station was on the seafront, specifically the stretch of water known as West Bay, with the water reaching right up to the southern edge of the platforms at high tide. A series of land reclamation projects to expand the docks, largely funded by the London & South Western Railway, culminated in the building of the vast 'New Docks' (now Western Docks) between 1927 and 1934, which led to all of West Bay being reclaimed and the station becoming landlocked. The new land and the demand for new lines allowed the station to be enlarged and redeveloped in 1934-1935 (from two platforms to four), and it became 'Southampton Central' on 7th July 1935. The new station buildings were largely constructed from concrete in the art deco style. An air raid on 23rd November 1940 damaged the buildings alongside platform one. The station was hit by two German parachute mines on 22nd July 1941, which destroyed the ticket hall on platform four and damaged the island platform. In preparation for the closure of Southampton Terminus station near the docks in 1966, alterations were made to the station's parcel handling facilities to allow it to handle increased volume. In 1967, soon after the closure of Southampton Terminus, the station was rebuilt, losing its clock tower that was replaced with an office block. On 10th July 1967 it was renamed 'Southampton', however, on 29th May 1994 it reverted to 'Southampton Central'. Of interest is the first carriage in the train. This was the only green-liveried Mark 2 first class corridor carriage to see service on the Southern – No S13389. *B. Wadey*

Above. No 80083 stands at Sandy station on 31st July 1954 with a motley collection of passenger coaches in tow. The locomotive was allocated new to Bletchley, where it remained – apart from a very short stay at Rugby – until reallocated to Neasden in September 1958. It was transferred to the Southern Region's Bricklayers Arms depot in December 1959. A final move to Eastleigh took place in June 1962 with it being withdrawn from there on 7th August 1966. The first section of the GNR – that from Louth to a junction with the Manchester, Sheffield & Lincolnshire Railway at Grimsby – opened on 1st March 1848, but the southern section of the main line, between Maiden Lane and Peterborough, was not opened until August 1850. Sandy was one of the original stations, opening with the line on 7th August 1850. The Sandy & Potton Railway was opened for goods traffic on 23rd June 1857, and to passengers on 9th November 1857. It was later purchased by the Bedford & Cambridge Railway (B&CR), which was absorbed by the LNWR in 1865. The eastern section of the Bedford-Cambridge route (sometimes known as the Varsity Line) closed on 1st January 1968, and with it, the ex-LNWR platforms at Sandy. The two stations were physically adjacent, and shared an island platform. In 1917 the LNWR station was placed under the management of the GNR, and then shared the booking facilities. After the closure of the Varsity Line, the station was considerably rebuilt in the early 1970s to give a four-track layout throughout, and platforms on the slow lines only, thus removing a two-track bottleneck on the East Coast main line. The first vehicle behind the locomotive is a horsebox, these wagons were vacuum braked and 'express' rated, enabling them to be used in passenger trains. *E. Sawford (1550)*

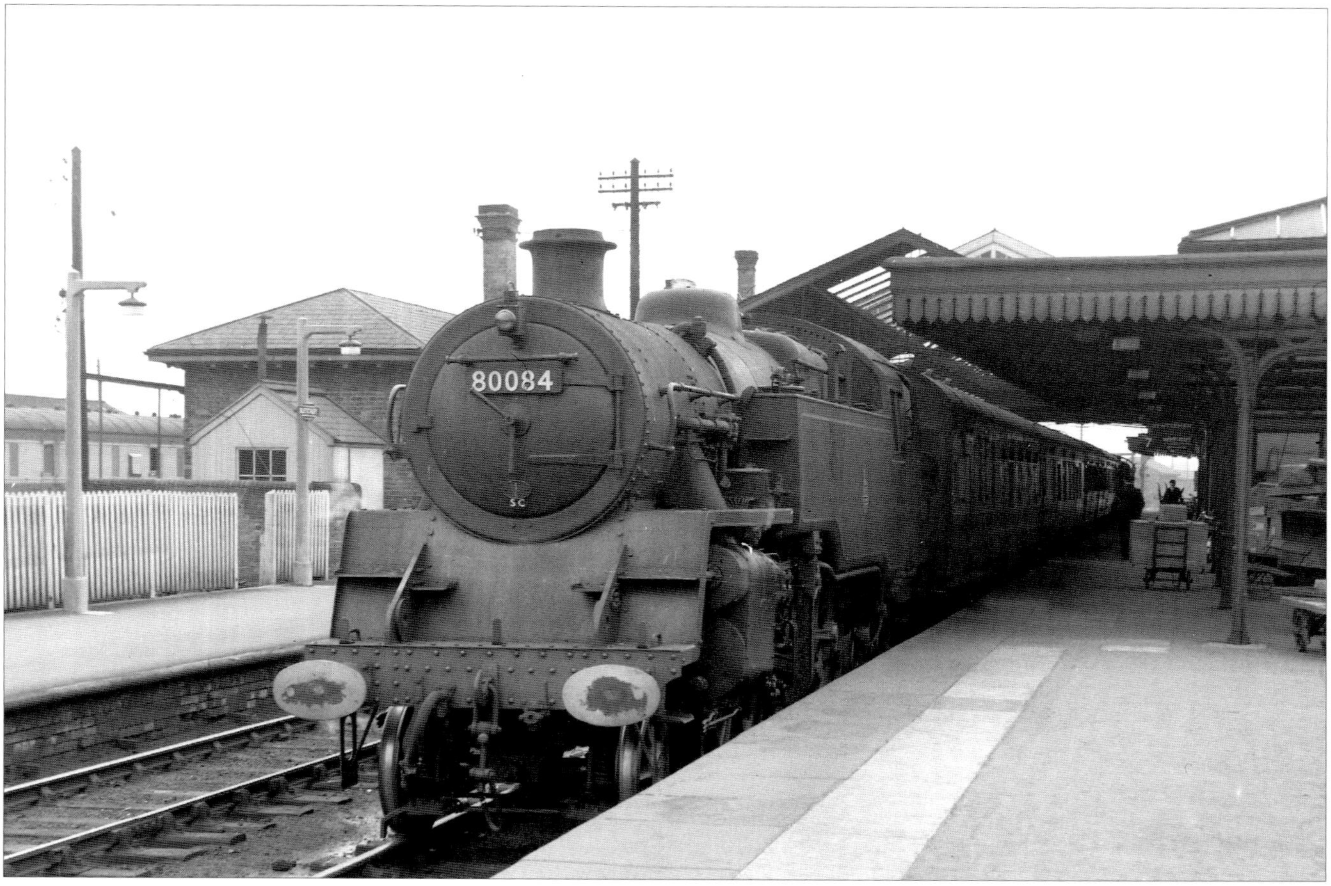
80084

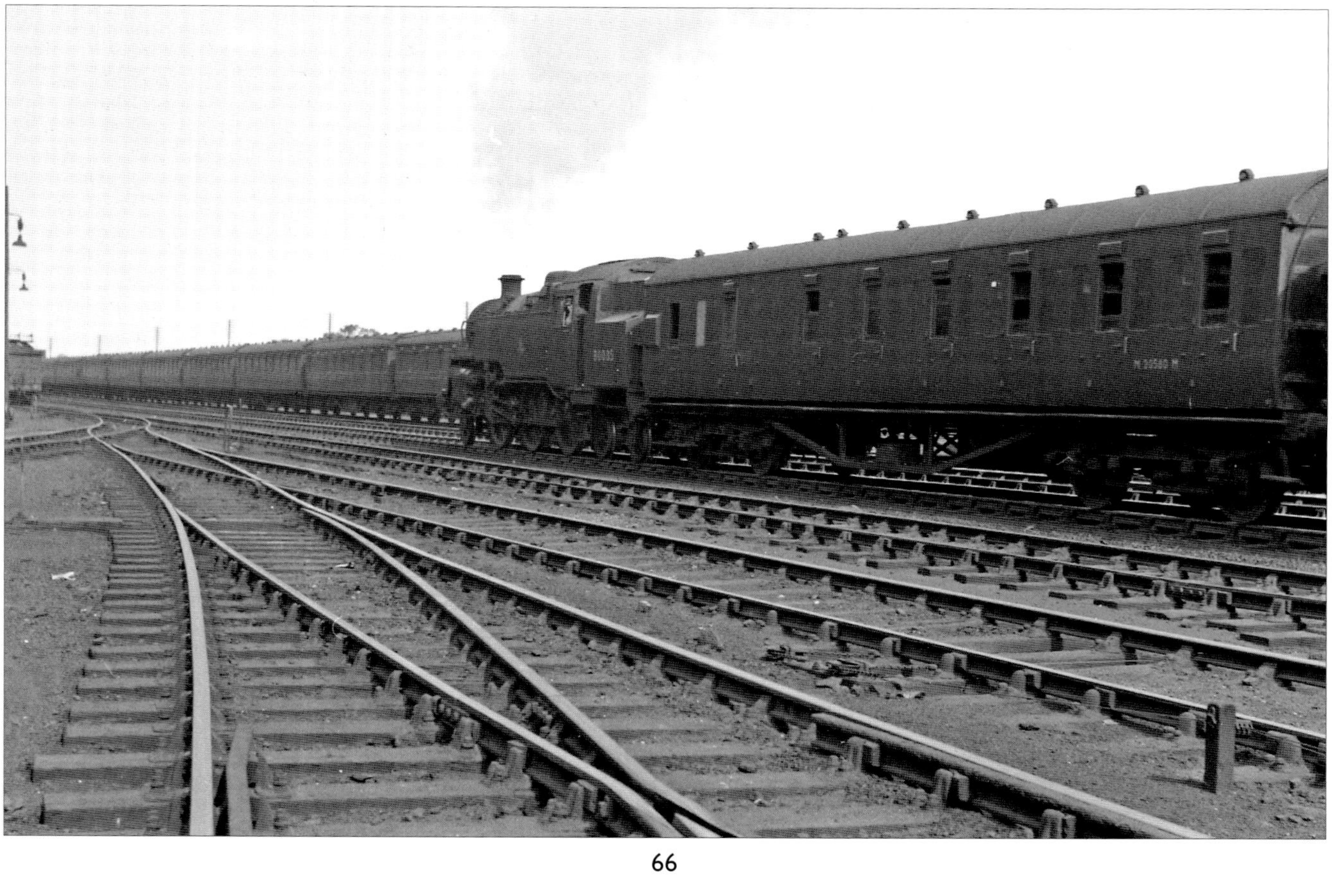

Top left. Allocated to the nearby depot, No 80084 is seen at Bletchley on 15th July 1954. The locomotive had entered service two months earlier on 14th May 1954. It was reallocated to Willesden in early November 1959 and moved south of the Thames in December that year, being transferred to Bricklayers Arms. It ended its days at Redhill on 13th June 1965, having seen duty at Stewarts Lane and Brighton. The London & Birmingham Railway, now part of the West Coast main line, was officially opened from Euston as far as Denbigh Hall (approximately one mile north of Bletchley station) on 9th April 1838, where a temporary station was built. The line was fully opened in September 1838, and Bletchley station opened some time between 2nd November 1838 and 20th June 1839. The station was known as Bletchley & Fenny Stratford between 1841 and 1846 and, after the opening of the Marston Vale line, was referred to in timetables as Bletchley Junction from 1851 to 1870. Originally a major intercity station, that role passed to Milton Keynes Central in 1982 when the latter was opened, long after the east-west route had been downgraded, taking Bletchley's importance as a junction with it. *E. Sawford (1477)*

Bottom left. A northbound local is leaving Bletchley behind No 80085 on 5th September 1954, four months after it entered traffic on 28th May. After a move to Rugby its days on the LMR ended in December 1959 when it moved to Bricklayers Arms. It moved around the Southern Region with allocations at Ashford, Tonbridge, Stewarts Lane, Brighton, Redhill and Feltham. A final move to Nine Elms in late summer 1966 resulted in it being withdrawn at the end of Southern Region steam on 9th July 1967. The leading carriage, No M20580M, is a Stanier-designed Non-Corridor Third Brake – dating from 1933, it was in traffic until November 1965. *E. Sawford (1678)*

Above. No 80088 was delivered new to Bury depot on 14th July 1954 with a move to North Wales coming in September 1956 when it was transferred to Bangor. Transfer to the Southern Region came in December 1959 with a move to Three Bridges. Reallocations to Tunbridge Wells West and Brighton saw it end its service life at Redhill on 13th June 1965. It is seen here taking water at Afon Wen during 1957. The station formed a junction between the Aberystwyth & Welsh Coast Railway (A&WCR) and the Caernarvonshire Railway, and opened to traffic on 2nd September 1867. Trains on the A&WCR line were operated by the Cambrian Railways, then absorbed into the Great Western Railway. Trains from the Caernarvonshire Railway were operated by the LNWR and so passed to the LMS. In addition to local services Afon Wen was served by trains from both London Paddington and London Euston. Those from Paddington would reach it on Cambrian rails through Machynlleth and Porthmadog, proceeding onward to terminate at Pwllheli. From Euston the train would travel via Crewe, Bangor and Caernarvon; at Afon Wen the front portion of the train would proceed forward to terminate at Porthmadog and the rear carriages would be detached for Pwllheli. The signal box and passing loop initially remained in use after the station closed on 7th December 1964, but after the lifting of the Caernarvon line, these were decommissioned in 1967 and removed three years later (leaving only the old westbound platform line in use as the running line to Pwllheli). Demolition of the surviving buildings and westbound platform followed by the late 1970s. *J. Flint/J. Harbart (FH1900)*

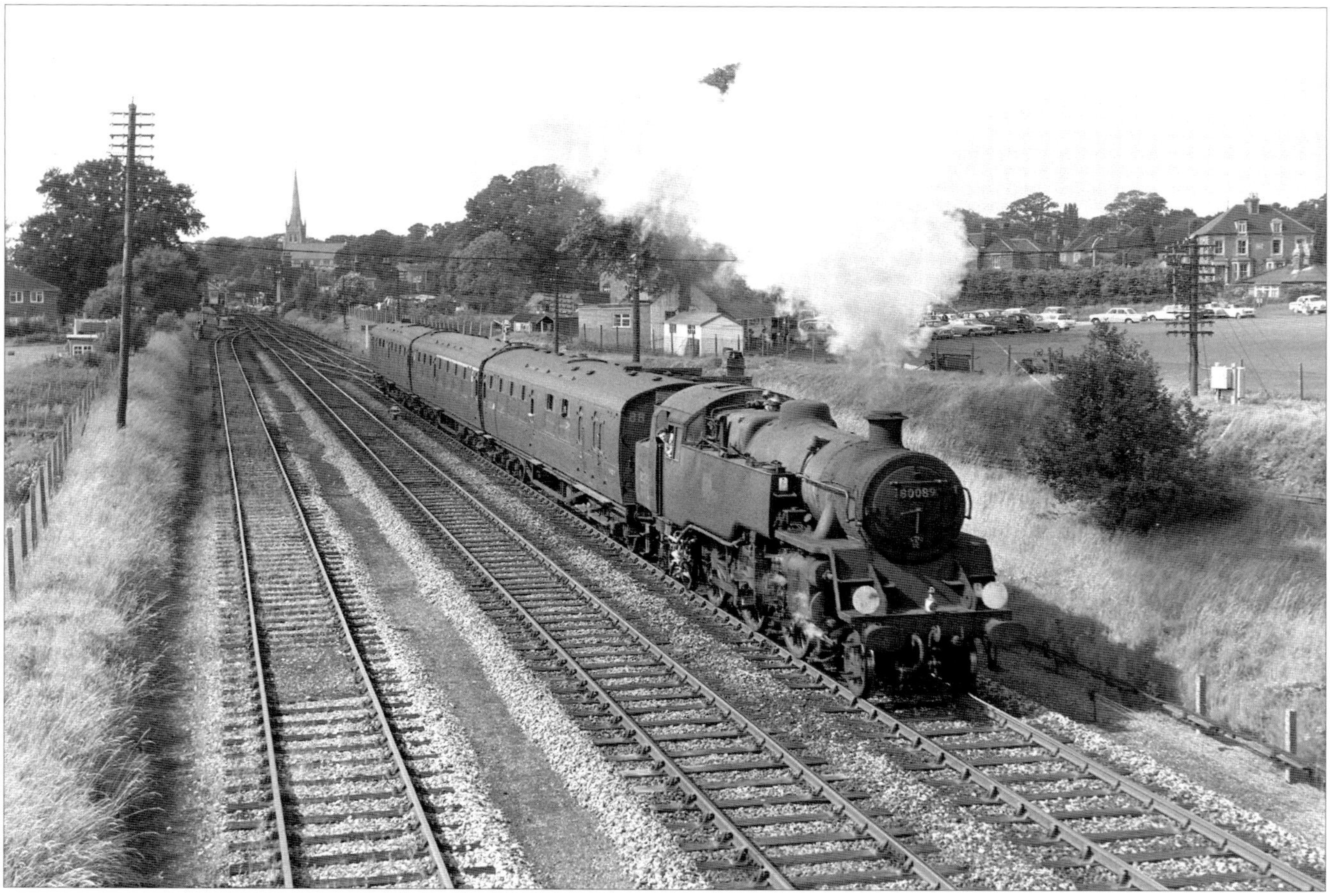
80089

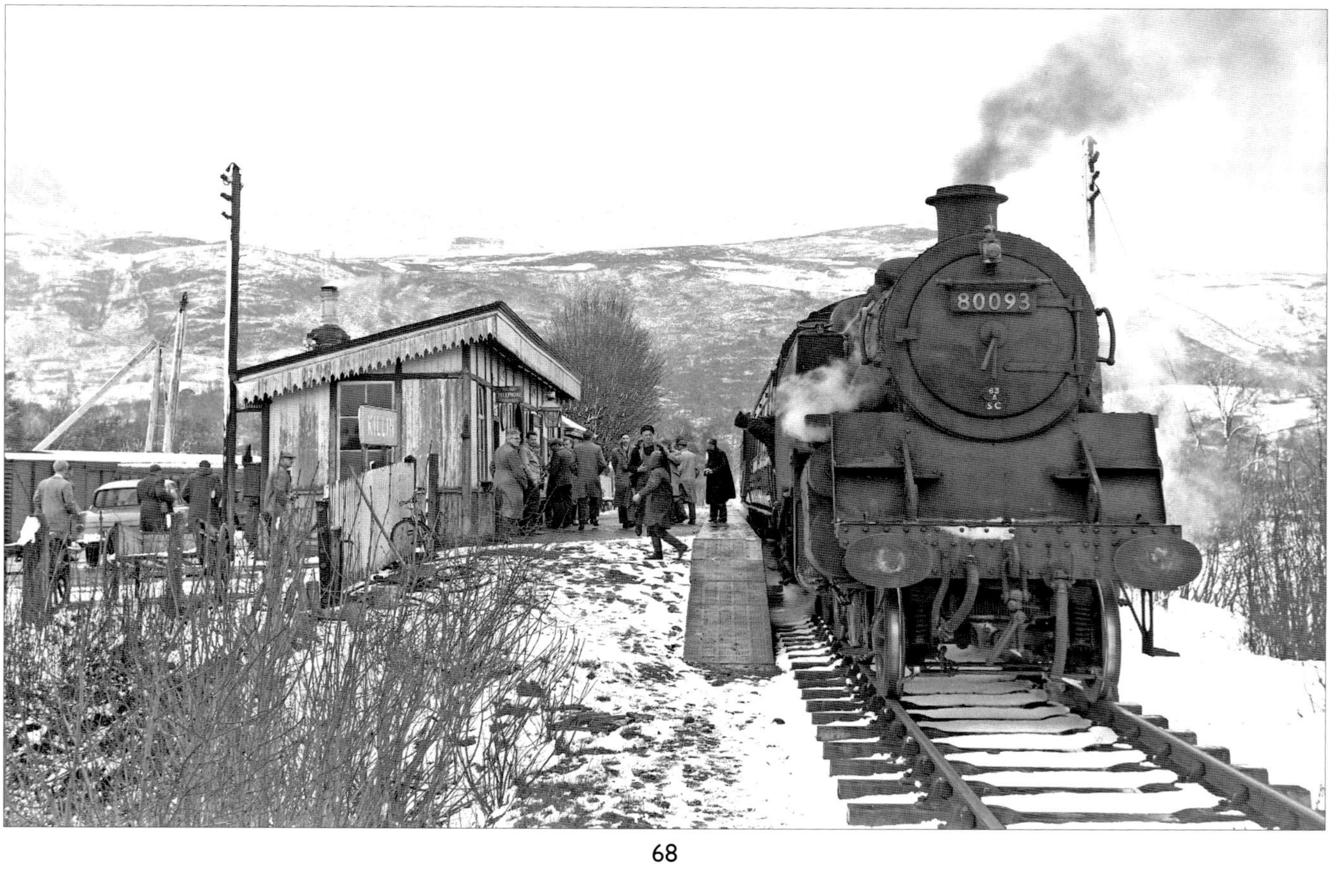
80093

Top left. Redhill allocated No 80089 is seen at the head of a local Reading-Redhill service departing Wokingham station on 20th June 1964. The locomotive had been allocated new to the LMR, being new to Bury on 3rd August 1954. It arrived at Three Bridges in December 1959 in the company of Nos 80087/88. It was reallocated to Stewarts Lane in January 1963, with later moves to Brighton, Redhill and Feltham. As with No 80087 it ended its days at Nine Elms, where it had arrived from Feltham just before Christmas 1965. It was withdrawn on 2nd October 1966 and was scrapped by Cashmore's, Newport, the following March. The line from Reading to Redhill was built by the Reading, Guildford & Reigate Railway (RG&RR), and was opened in stages. The first sections, from Reading to Farnborough North, which included a station at Wokingham, also from Dorking West to Redhill, were opened on 4th July 1849. Other sections followed, with the last section, from Guildford (Surrey) to Shalford, on 20th October 1849. From its beginning the RG&RR was worked by the South Eastern Railway (SER), which bought the RG&RR in 1852. The Staines, Wokingham & Woking Junction Railway (SW&WJR) opened a line between Staines and Wokingham (Staines Junction) on 9th July 1856. The LSWR worked the SW&WJR and was authorised to run over the SER to Reading. This gave Wokingham a direct route to London Waterloo. *B Wadey*

Bottom left. No 80093 was new to Bedford on 11th October 1954, moving around the LMR with allocations at Bury, Newton Heath, and Blackpool Central before its arrival at Perth South depot in spring 1960. It is seen here at Killin with the Stephenson Locomotive Society (Scottish Branch) and the Branch Line Society's 'Scottish Rambler No 2 (Joint) Easter Rail Tour on 12th April 1963. Spring in Scotland was late that year! No 80093 worked the Killin Junction-Killin-Loch Tay, and return, section of the rail tour. The tour was over four days (12th-15th). No 80093 was withdrawn on 26th September 1966, meeting its fate at the hands of Cambell's cutters in its Airdrie scrapyard. The Killin Railway was a locally promoted line built to connect the town of Killin to the Callander & Oban Railway main line nearby. It opened in 1886, and carried tourist traffic for steamers on Loch Tay as well as local business. The directors and the majority of the shareholders were local people, and the little company retained its independence until 1923. Goods service on the branch was discontinued on 7th November 1964. The C&OR line was to be closed completely on 1st November 1965, and the Killin branch with it. In fact there was a serious rock fall on the C&OR line in Glen Ogle on 27th September 1965, and the line was impassable; clearing the line was unaffordable, and the line never re-opened; the Killin branch closed prematurely the next day. *D. Clarke*

Above. Despite most of the images in this volume show members of the class hauling passenger traffic they were just at home on freight workings. No 80094 is seen during its time allocated to Bangor shunting a rake of mineral wagons in what is believed to be the goods yard at Bethesda. The locomotive was new to Kentish Town on 25th October 1954, being transferred to Bangor in late summer 1956. Three years late it arrived at Birkenhead Mollington Street, leaving there for the Southern Region at Three Bridges in December 1959. As with other LMR locomotives sent south it saw service at Stewarts Lane, Brighton, Redhill and Feltham. It was withdrawn from the latter on 31st July 1966 with a move to South Wales for scrapping at Cohen's Morriston yard. *J. Flint/J. Harbart (FH2572)*

Top left. No 80095 is seen resting between duties on Bangor shed during 1957. It was new to St Albans on 9th November 1954, moving to North Wales in September 1956. A final move during its LMR days was to Birkenhead Mollington Street where it arrived in October 1959. That December saw its reallocation to Tunbridge Wells West, subsequent moves around the Southern Region saw it allocated to Eastleigh, Guildford and Feltham. A final move to Nine Elms occurred in November 1964 from where it was withdrawn on 2nd October 1966. The first depot at Bangor was opened by the Chester & Holyhead Railway on 1st May 1848. Absorbed by the LNWR, it was closed on 1st January 1859. It was replaced by an 1859 LNWR shed, which in turn was replaced in 1884. The shed was reroofed by BR in 1957 and closed on 14th June 1965. *J. Flint/J. Harbart (FH2573)*

Bottom left. Missing its shed and self-cleaning plates on the smoke box door, No 80096 is seen on shed at Plaistow in 1959. The locomotive was reallocated away that November when the shed closed so may well have had them removed in anticipation of the transfer. It was new from Brighton on 23rd November 1954, and spent five years there prior to its move to Tilbury. It was moved to the Western Region on 15th July 1962 when it was transferred to Shrewsbury. It was reallocated to Croes Newydd in early November before a move to Machynlleth in early March 1963. That September saw a return to Croes Newydd, by which date the shed was part of the LMR. A final move to Bournemouth took place in late April 1965 from where it was withdrawn on 26th December 1965. It was scrapped locally by T W. Ward in Ringwood goods yard during the following March. The original six-road Plaistow shed was opened by the LT&SR in 1896, closing in 1911. Its replacement had eight roads and opened the same year, closing on 2nd November 1959 when its allocation of Standard Fours moved to Tilbury.

Above. Having joined the exodus from Plaistow, by June 1963 No 80097 found itself allocated to Stratford. New to traffic on 9th December 1954, it is seen here on 3rd July 1962 acting as station pilot at Liverpool Street, a few days before it move to the Western Region at Swansea East Dock which took place on 15th July. It moved north to Oswestry on 14th July 1963 before a final move to Machynlleth in late June 1964. Withdrawn on 24th July 1965 it was sold to Woodham Bros at Barry and subsequently preserved.

Top left. Doncaster Works constructed the 80106-115 series with delivery to traffic commencing on 22nd October 1954, with all 10 locomotives destined for the Scottish Region. No 80108 was new to traffic on 5th November 1954 and is seen here on the 7th waiting its trip to Kittybrewster. Noticeable is the speedometer drive from the trailing driving wheel, and the use of larger numerals on the bunker. Note that the power class rating is missing and the LNER/ER route availability rating. No 80108 departed Kittybrewster in late spring 1957 destined for Polmadie from where it was withdrawn on 8th May 1965.

Bottom left. The location is obvious as No 80110 reverses past Renfrew South No 2 signal box on 21st June 1962 – the photographer did not note the number of the locomotive on the leading end of the train. The locomotive was new to Kittybrewster on 18th November 1954, moving to Polmadie in June 1957 – where it was allocated at the time of the photograph. It was transferred to Carstairs in late summer 1963 before a return to Polmadie in May 1964. It was withdrawn on 8th May 1965 and sold to Motherwell Machinery & Scrap, Wishaw, for recycling. South Renfrew station opened on 19th April 1897 by the Glasgow & South Western Railway. To the west was Renfrew Steel Works and to the east was Albert Cabinet Works. The steel works was served by a signal box called Porterfield, that was to the south and it opened before the station. Two replacement signal boxes were built when the station opened, which were named 'Renfrew South No 1' and 'Renfrew South No 2'. Also to the south was a yard that served the sidings for the works. The station closed on 5th June 1967. *W. A. C. Smith (6141)*

Above. New from Doncaster on 3rd December 1954, No 80112 was allocated to Polmadie where it stayed until transferred to Kittybrewster in late spring 1957. The summer of 1961 saw it reallocated to Hurlford, a final move to Corkerhill took place in the summer of 1964. It is seen a year later on 29th July 1965 at Crossmyloof with the 5.08pm Glasgow St Enoch-East Kilbride service. The station serves the areas of Crossmyloof and Shawlands in Glasgow, being located 1mile 60 chains (2.8km) from Glasgow Central. The station was opened on 1st June 1888 by the Glasgow, Barrhead & Kilmarnock Joint Railway that was jointly owned by the Caledonian Railway and the Glasgow & South Western Railway, giving the latter a shorter access to its Carlisle main line. The station was renovated in the 1990s, during which an over-line station building was demolished. *J. L. Stevenson*

Delivered to traffic from Doncaster Works on 14th December 1954 No 80113 made its way to Polmadie where it was to remain until June 1957 when it was reallocated to Kittybrewster. A move to Keith took place in June 1960, and a year later saw it allocated to Hawick. A final move to St Margarets occurred early in 1966 from where it was withdrawn on 2nd September 1966. Note the recess in the cab side that was used to house single-line tablet-catcher required on some routes, chiefly those of the former GNSR. The shed at Hawick was opened by the Edinburgh & Hawick Railway on 1st November 1849, being sited to the north of the station. It was reroofed by the Scottish Region in 1955 and closed on 3rd January 1966.

# The Class of 1954

The final Class Four to be delivered in 1954 was No 80115 that was delivered from Doncaster Works on 31st December, destined for Polmadie. It is seen on 2nd April 1955 still in remarkably clean condition. A bit of a nomad for this batch saw it allocated to Kittybrewster, Keith and Aberdeen Ferryhill, before a return to Polmadie in August 1961. Three years later in January 1964 saw a move to Ardrossan where it stayed for six months before a final return to Polmadie in July from where it was withdrawn on 12th October. It joined many others of the class being scrapped by Motherwell Machinery & Scrap's cutters at Wishaw.

# British Railways Standard Tanks

Below. No 82020 stands at Dovey Junction on 23rd July 1963 with the 4.50pm Pwllheli-Machynlleth service. Delivered new to Hull (Botanic Gardens) from Swindon Works on 29th September 1954, it was reallocated to the LMR a few weeks later when it arrived at Nuneaton, initially on loan, the transfer was later made permanent. Two years later it was at Wrexham Rhosddu, before a move to the Western Region at Shrewsbury in January 1960. By March it was at Machynlleth where it remained until a transfer to the Southern Region in spring 1965. It was withdrawn from Nine Elms depot on 19th September 1965, destined for the scrap pile at Birds, Risca. The station was opened in 1863 as Glandovey Junction by the Aberystwyth & Welsh Coast Railway, being where the route splits with lines to Aberystwyth and Pwllheli. From 1865 it became part of the Cambrian Railways and was renamed Dovey Junction on 1st July 1904. *B. Wadey*

Top right. No 82021 arrived at Hull (Botanic Gardens) following delivery from Swindon Works on 6th October 1954. A similar journey to No 82020 was taken which saw it arrive at Machynlleth in spring 1960. It is seen here on a shed that was opened by the Newtown & Machynlleth Railway on 3rd January 1863, the Cambrian Railways absorbed the facilities on 25th July 1904. At a later date the shed was extended and ultimately closed to steam on 5th December 1965, remaining open for the servicing of DMUs. No 82021 was not there at the end, as it had moved to Nine Elms with No 82020. It was withdrawn on 17th October 1965, after a service life of a few days over 11 years.

Bottom right. Resting between duties on Nine Elms depot is No 82022, with classmate 82019 also in shot. It was allocated new to Exmouth Junction on 15th October 1954 having been delivered from Swindon Works. It stayed in Devon until September 1962 when it was reallocated to Eastleigh. A final move to Nine Elms occurred that November, from where it was withdrawn on 17th October 1965. It probably made the journey to Buttigiegs, Newport, with No 82021. Also in view are two rebuilt Bulleid Pacifics and an air-smoothed lightweight version that appears to be No 34070 *Manston. J. Flint/J. Harbart (FH1901)*

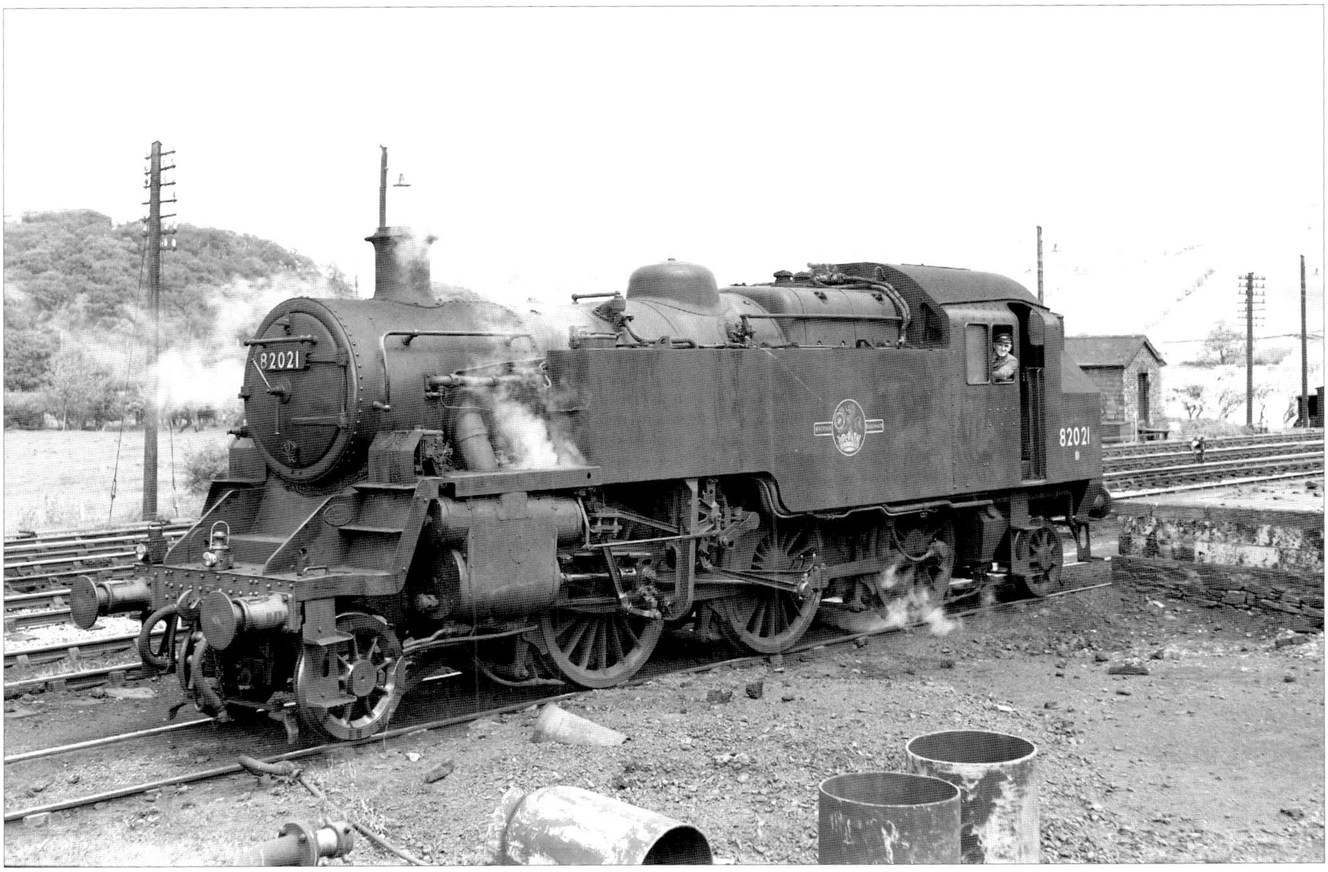
82021
82021

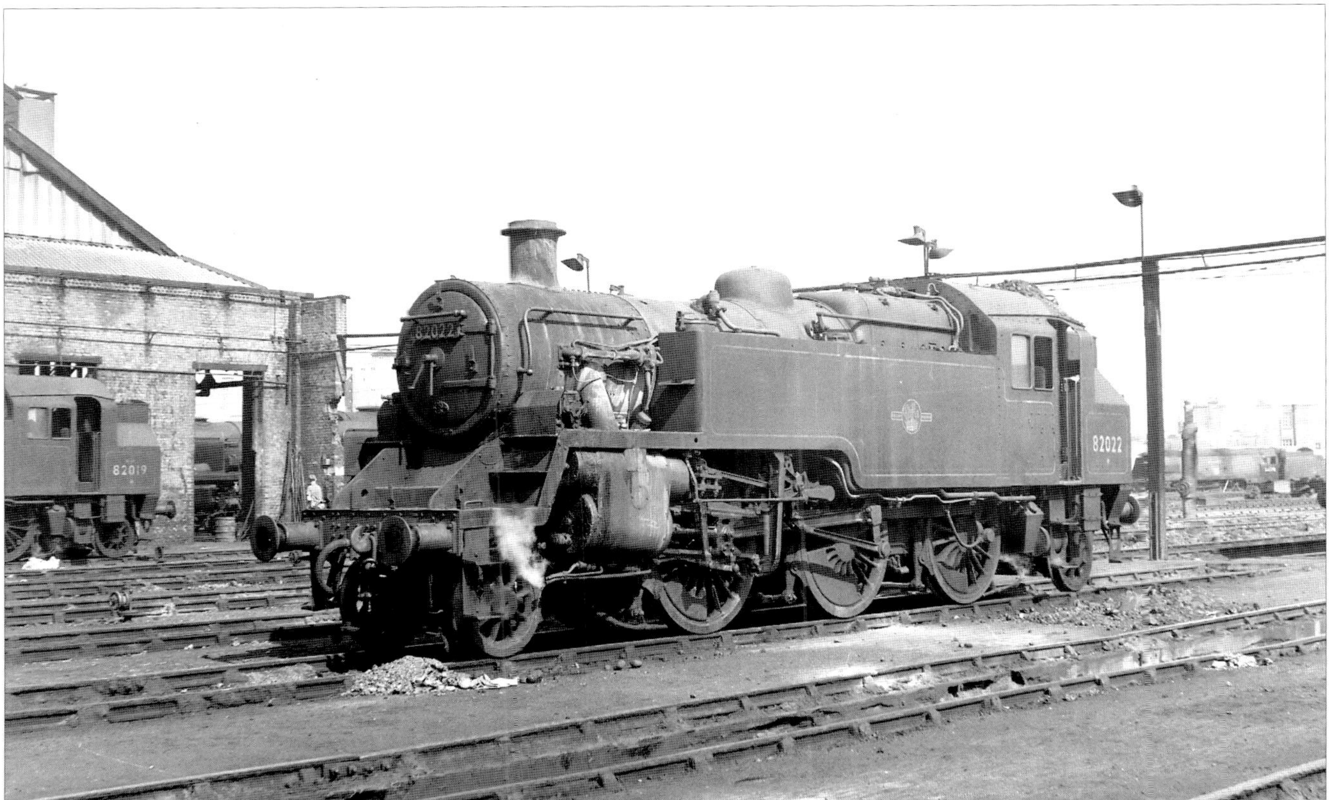
82022
82019
82022

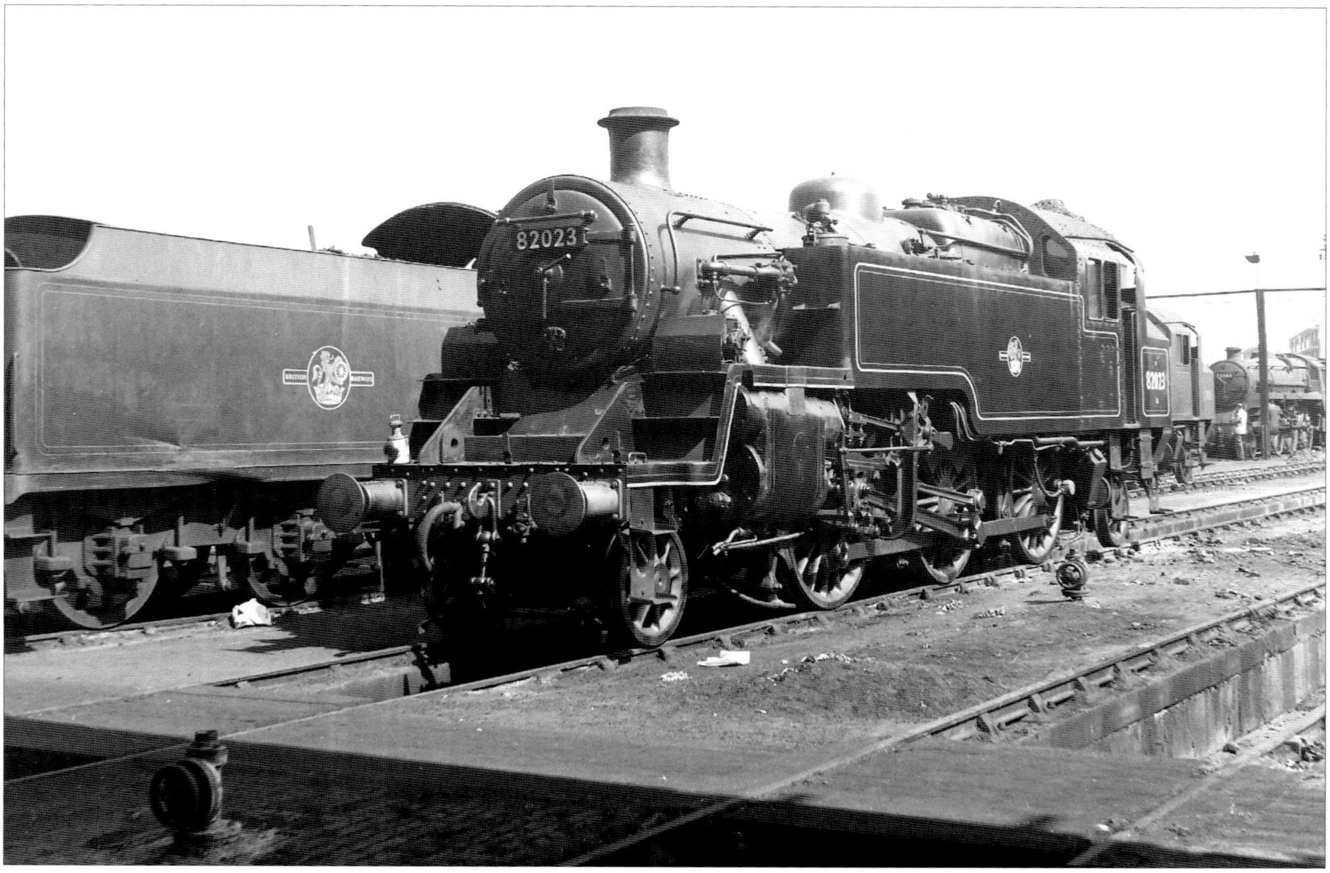
82023
82023

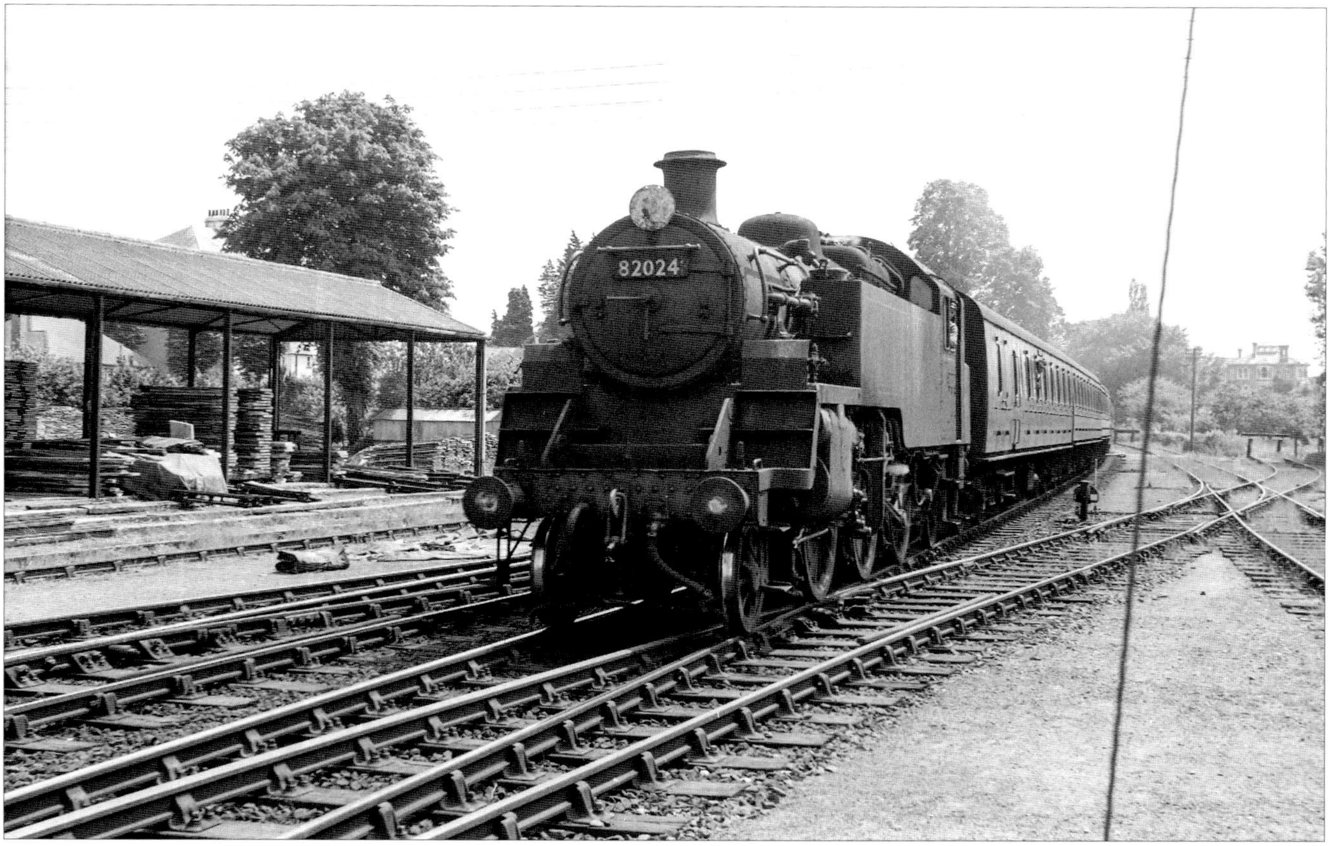
82024

Top left. A smart looking No 82023 on shed at Nine Elms in company with a classmate and Standard Class Five No 73088 *Joyous Gard*. New from Swindon Works on 22nd October 1954, the locomotive was allocated to Exmouth Junction, where it remained until September 1962 when it was reallocated to Eastleigh. A final move that November saw it at Nine Elms from where it was withdrawn on 2nd October 1966. The tender of the unidentified Schools class locomotive alongside appears to have been sideswiped at some time. *J. Flint/J. Harbart FH(1902)*

Bottom left. New to Exmouth Junction on 29th October 1954, No 82024 is seen here with the 12.45pm service from Exmouth entering Topsham station on 25th May 1961. The station, with buildings designed by Sir William Tite, were opened by the London & South Western Railway on 1st May 1861. The siding off to the right led down to the quayside on the River Exe, the 32-chain line was opened to traffic on 23rd September 1861. One of the important commodities unloaded at Topsham was guano that was lifted out of ships' holds in baskets. An engine went down to the Quay in the morning to collect wagons that were then worked from Topsham goods yard to Odam's siding by the afternoon goods. Barrels came by rail to the quay from Peterhead for sprats to be packed in and were then loaded into the Tuborg lager boats to form a cargo for their return journey. Bundles of half-pound boxes of smoked sprats were sent off by rail from the quay. Latterly the Quay branch was worked three times a week until its closure in 1957, rails and sleepers being lifted in August the following year. No 82024 was to end its career at Nine Elms on 30th January 1966.

Above. Inside Swindon Works in the autumn of 1954 Nos 82025, 82027 and 82028 are seen in various stages of construction. No 82025 was to enter service on 5th November 1954, being allocated to Exmouth Junction. Following the by now familiar route to Nine Elms it was withdrawn on 9th August 1964 after less than 10 years in traffic. Of note is the use of 'dollies' to hold the pony trucks square enabling them to be moved easily under the locomotives when the time came.

Top left. No 82026 entered service from Swindon Works on 12th November 1954, being allocated to Kirkby Stephen. It moved around the North Eastern Region with stays at Darlington, Scarborough, Low Moor, Copley Hill (Leeds) before heading south to Guildford in September 1963. Following a move to Bournemouth it is seen at Ringwood with the 11.04am Brockenhurst-Bournemouth Central service on 25th April 1964. A final move to Nine Elms saw it being withdrawn from service on 26th June 1966. The Ringwood, Christchurch & Bournemouth Railway was formed to link Christchurch and Bournemouth, to the London & South Western Railway's Southampton and Dorchester line at Ringwood; the station opened on 1st June 1847. The RC&BR opened in 1862 from Christchurch to Ringwood, and was extended to Bournemouth in 1870, sharing in the growing popularity of the town. However the route was circuitous – becoming known as Castleman's Corkscrew – and the LSWR opened a shorter route between Brockenhurst and Christchurch via Sway in 1888, making the Ringwood to Christchurch section a branch line. The station was closed to passengers on 4th May 1964 and to freight on 7th August 1967. The station was demolished after closure and most of the site has been redeveloped with industrial units. The trackbed approaching the town from each direction is now part of the Castleman Trailway.

Bottom left. No 82027 is seen taking water at the buffer stops at Waterloo station, having arrived with an empty coaching stock working from Clapham Junction storage sidings. It had been new to Kirkby Stephen on 23rd November 1954, moving around the region with periods at West Hartlepool, Malton, Scarborough and York (North) before heading south in September 1963. Following a stay at Bournemouth, with Nos 82026/28/29 it accompanied them to Nine Elms from where it was withdrawn on 9th January 1966.

Above. No 82028 is seen passing Bog Hall (Whitby) signal box with a three coach local service in 1962 whilst allocated to Malton depot where it had arrived from Scarborough in September 1961. It had been new to Darlington on 2nd December 1954, and departed from York (North) for the south in September 1963. As with its classmates it ended its days at Nine Elms on 4th September 1966. Whitby's original railway station stood near to the end of the current platform, in the form of the offices, workshop and carriage shed of the Whitby & Pickering Railway; a single track horse worked line opened throughout in 1836. In 1845, the W&PR was taken over by the York & North Midland Railway and converted into a double tracked, steam worked line. The Y&NMR built the present Whitby station to the design of its architect George Townsend Andrews, the station included a fine 'Euston Truss' overall roof that was removed by BR in 1953 and replaced by the present awnings. In 1854, the Y&NMR helped form the North Eastern Railway, who later added two more platforms to help deal with traffic from the other branch lines that served Whitby; the Esk Valley Line finally opened throughout to a junction at Grosmont in 1865 while the coast line from Loftus opened in 1883 and from Scarborough in 1885. The iconic abbey ruins can be seen atop the hill. *N. Stead collection (NS201888)*

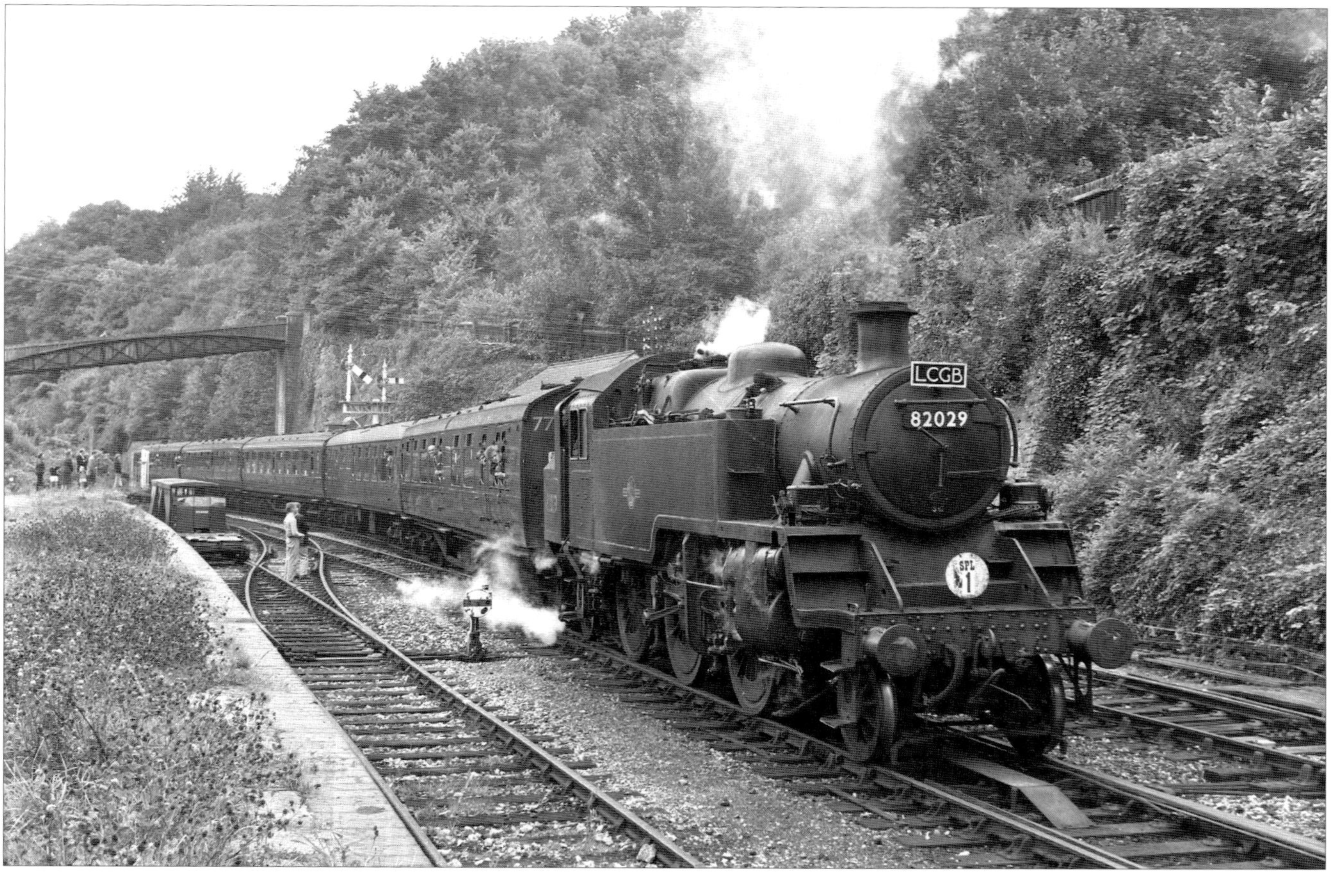
LCGB
82029
1

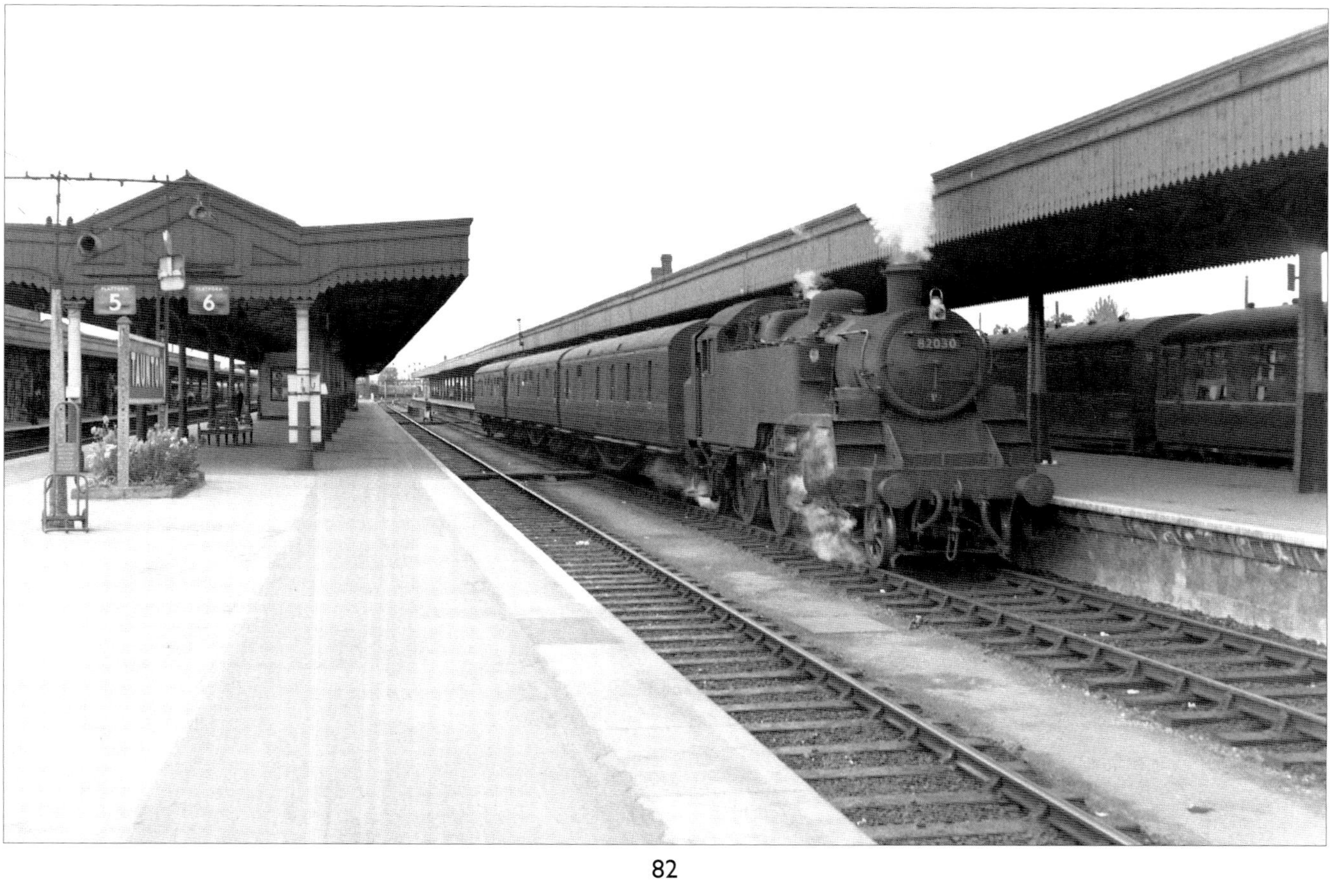
5
6
82030

Top left. The Locomotive Club of Great Britain's 'Anton & Test Valley Railtour' of 6th September 1964 is seen departing the former GWR station of Winchester (Chesil), behind No 82029 during its time allocated to Bournemouth. New to Darlington on 13th December 1954, a final move to Nine Elms saw it withdrawn on 9th July 1967 at the end of steam on the Southern. The tour commenced at Winchester (Chesil) and, after visiting the Ludgershall branch, ended at Eastleigh. The station, as Winchester Cheesehill, was opened on 4th May 1885 and for the first six years after the opening of the line was the terminus of the Didcot, Newbury & Southampton Railway (DNSR), until the line was extended to link up with the Southern Railway line to Southampton. The station buildings were larger than those of any other DNSR station but were built to the standard designs used by the GWR. The station buildings were located on the northbound platform. At the northern end of the station the line passed into the double tracked Chesil Tunnel. The station closed temporarily on 4th August 1942, reopening on 8th March 1943. The station was renamed Winchester Chesil on 26th September 1949. Like the other stations on the southern part of the line, Winchester Chesil closed on 7th March 1960; but unlike the others, it was reopened for the next two summers: 18th June 1960 to 10th September 1960 and 17th June 1961 to 9th September 1961, on Saturdays only. *B. Wadey*

Bottom left. No 82030 stands in platform 7 at Taunton in June 1964, shortly before being reallocated to Exmouth Junction – that was by then part of the Western Region. It had been new from Swindon Works on 21st December 1954, being allocated to Barry. It was shunted around the region on a regular basis before arriving at Taunton with periods at Shrewsbury, Worcester, Bristol Bath Road, Kidderminster, Wellington and Bristol Barrow Road. It left Exmouth Junction in the early summer of 1965 when it was allocated to Gloucester Horton Road from where it was initially withdrawn on 2nd August 1965. Following a period in store it was reinstated, on 22nd November 1965, and sent to Bath Green Park from where it was withdrawn for a second time on 31st December 1965. Originally opened on 1st July 1842 as part of the Bristol & Exeter Railway, Taunton was the terminus of the line until a new temporary terminus was opened on 1st May 1843 further west at Beambridge. A series of branches opened in the area during the next 30 years. These were the Yeovil branch (1st October 1853), the West Somerset Railway to Watchet (31st March 1862), the Chard branch (11th September 1866), and the Devon & Somerset Railway (8th June 1871, extended to Barnstaple 1st November 1873). While none of these branches had a junction in Taunton, the trains were generally run through to Taunton to provide connections. *W. A. C. Smith (7471)*

Above. The final Class 3 delivery in 1954 was No 82031 that was new to Barry on 30th December. Following periods at Laira, Newton Abbot, Shrewsbury and Wrexham Rhosddu it arrived at Machynlleth in January 1960. It is seen here on the final approaches to Pwllheli. In 1861 the Aberystwyth & Welsh Coast Railway was given authorisation to build a line along Cardigan Bay between Aberystwyth and Porthdinllaen on the Llŷn Peninsula. However, the final five miles across the Llŷn Peninsula were never built. By 1865 the company had merged to become part of the Cambrian Railways. When the first Pwllheli station opened on 10th October 1867 the decision to not complete the final five miles to Nefyn had already been taken. The station, which was about a half a mile from the town, became the line's terminus. On 19th July 1909 a second station was opened nearby the town centre following land reclamation that permitted the extension of the line. It had two tracks separated by an island platform with a small loading dock to the north. The layout remained unchanged until rationalisation began in September 1977. Prior to the closure of the Afon Wen to Caernarfon line in 1964, there were two named daily express services during the summer between Pwllheli and London: the 'Cambrian Coast Express' ran via Machynlleth, Shrewsbury and Birmingham to London Paddington and 'The Welshman' ran via Caernarfon and Crewe to Euston. No 82031 left the Western Region in June 1964 when it departed for the London Midland Region at Bangor. A final move to Patricroft in late March/early April 1965 resulted in its withdrawal on 10th December 1966, a few days short of 12 years in traffic. It returned to South Wales for disposal. *E. N. Kneale (K842)*

# The Class of 1955

Top left. The last of the 80054-58 series, No 80058, was new from Derby Works on 8th January 1955, and was sent north to Polmadie; that was to be the only depot that the locomotive was ever allocated to. It is seen here in October 1964 between Shields Junction and Paisley Gilmour Street. It was withdrawn on 17th July 1966 and consigned to Shipbreaking Industries at Faslane for disposal. This section of line between Glasgow Bridge Street station and Paisley was owned by the Glasgow & Paisley Joint Railway. It was constructed and operated jointly by two competing companies as the stem of their lines to Greenock and Ayr respectively, and it opened in 1840. The Joint Committee, which controlled the line, built a branch to Govan and later to Cessnock Dock, and then Prince's Dock.
*W. A. C. Smith (7811)*

Bottom left. Delivered new from Brighton Works on 17th January 1955, No 80099 is seen departing from Leigh-on-Sea station with a service to Southend and Shoeburyness. At this date the locomotive was allocated to Plaistow, it was reallocated to Tilbury in November 1959 with a further move to Stratford in June 1962. It was reallocated to the Western Region on 15th July the same year when it was sent west to Swansea East Dock. On 14th July 1963 it was sent to Machynlleth from where it was withdrawn on 8th May 1965, heading back to South Wales for recycling. The station was originally opened as Leigh by the London, Tilbury & Southend Railway on 1st July 1855, being renamed Leigh-on-Sea on 1st October 1904, but was rebuilt by the LMS on a new site 880yd (805m) to the west, opening on 1st January 1934.

Above. What was recoded as the Pwllheli portion of the Up 'Cambrian Coast Express' is seen here on 30th July 1963 at Porthmadoc behind No 80101 – minus any identification on the smoke box door. The station was opened by the Aberystwith & Welsh Coast Railway as Portmadoc on 12th September 1867, and renamed Porthmadog or 5th May 1975. No 80101 had been new to Plaistow on 7th February 1955 and had left the LT&SR's Tilbury depot in June 1962 for Stratford. A move to the Western Region's Shrewsbury depot came on 17th July 1962, moving west within two months to Machynlleth where it would stay until February 1963 when it moved to Croes Newydd. A move back to 'Mac' followed in March, which ultimately saw its withdrawal from there on 17th July 1965. *B. Wadey*

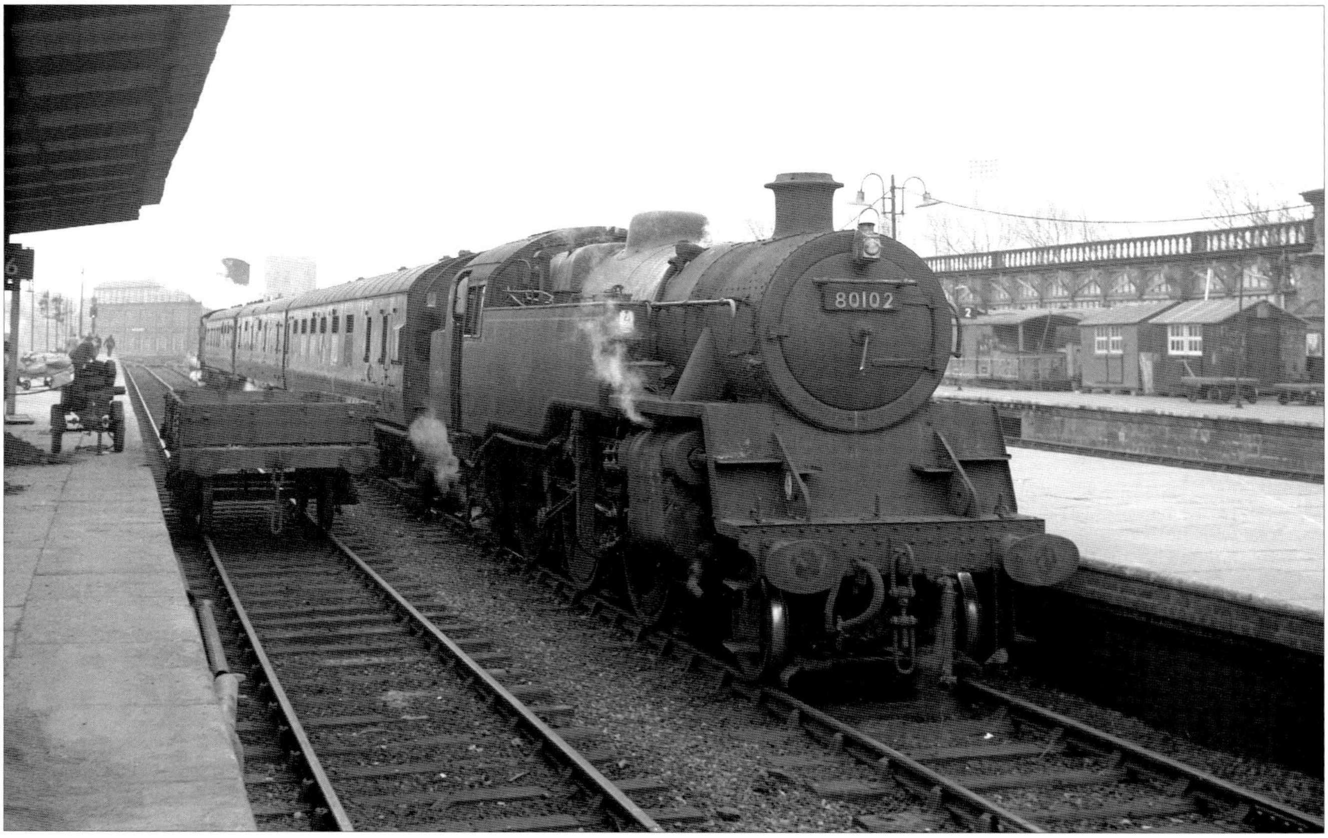
80102

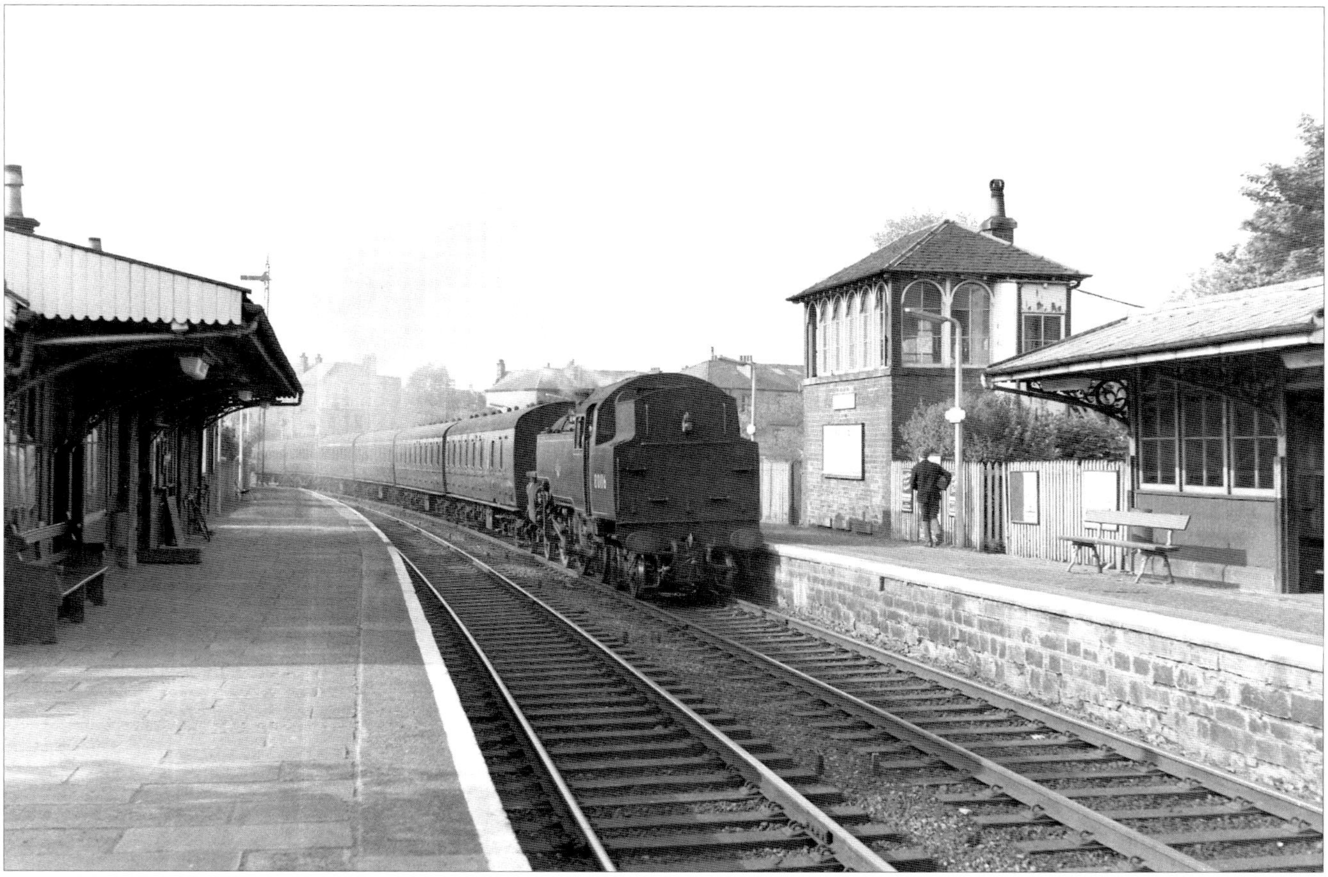

Top left. On 10th April 1964 the 5.30pm service from Welshpool has arrived at Shrewsbury behird No 80102, by the end of June the locomotive would be reallocated to Bangor, before returning in September. The locomotive had been new to Plaistow on 1st March 1955, moving to Tilbury before transfer to the Western Region's Old Oak Common depot on 15th July 1962, moving north to Shrewsbury before the end of the year. Shrewsbury's station was originally built in October 1848 for Shropshire's first railway — the Shrewsbury to Chester line. The station was extended between 1899 and 1903 by the construction of a new floor underneath the original station building. The building style was imitation Tudor, complete with carvings of Tudor style heads around the window frames. This was done to match the Tudor building of Shrewsbury School (now Shrewsbury Library) almost directly opposite and uphill from the station. The station's platforms also extended over the River Severn. It was operated jointly by the GWR and the LNWR.
*J. L. Stevenson*

Bottom left. No 80116 was delivered from Brighton Works and allocated to the North Eastern Region's depot at York (North) on 4th May 1955. Following periods at Whitby, Scarborough, Neville Hill and Holbeck it was reallocated to the Scottish Region in autumn 1963 when it moved to Dumfries. By the end of November it was at Carstairs with a final move to Polmadie in May 1964. It is seen here arriving at Giffnock with the 5.33pm service from St Enoch to East Kilbride on 28th May 1964. It was withdrawn from Polmadie on 1st May 1967: the date of closure of Polmadie to steam, No 80116 having been the last steam station pilot in Glasgow Central. It was dispatched to Campbell's Ardrie depot for disposal. The station at Giffnock was opened by the Busby Railway on 1st January 1866, this being a line built to the south of Glasgow, connecting the (at the time) small villages of Thornliebank, Giffnock, Clarkston and Busby and later Thorntonhall and East Kilbride with the city. It opened in two stages, in 1866 and 1868, and served industry and encouraged residential development.*W. A. C. Smith (7458)*

Above. The 8am service from Stranraer is seen at New Galloway on 11th June 1965 behind No 80117. New to Whitby on 19th May 1955 the locomotive arrived in Scotland in October 1963, being allocated to Dumfries, moving to Beattock in November 1965 and then to Polmadie a few weeks later. Withdrawn on 3rd March 1966, it was sent for scrap at Wishaw by the Motherwell Machinery & Scrap company. New Galloway station served the town in the historic county of Kirkcudbrightshire, being opened on 12th March 1861 by the Portpatrick Railway (PPR). The line was later operated by the Portpatrick & Wigtownshire Joint Railway (P&WJR). The title appeared in 1885 when the previously independent PPR and Wigtownshire Railway companies were amalgamated by Act of Parliament into a new company jointly owned by the Caledonian Railway, Glasgow & South Western Railway, Midland Railway and the LNWR and managed by a committee, the P&WJR. The PPR route, often known as the Port Road, linked Dumfries, via Castle Douglas, with the port towns of Portpatrick and Stranraer. It also formed part of a route by rail and sea from England and Scotland to the north of Ireland. The line and station closed to both passengers and goods on 14th June 1965 as part of Dr Beeching's axe. *J. L. Stevenson*

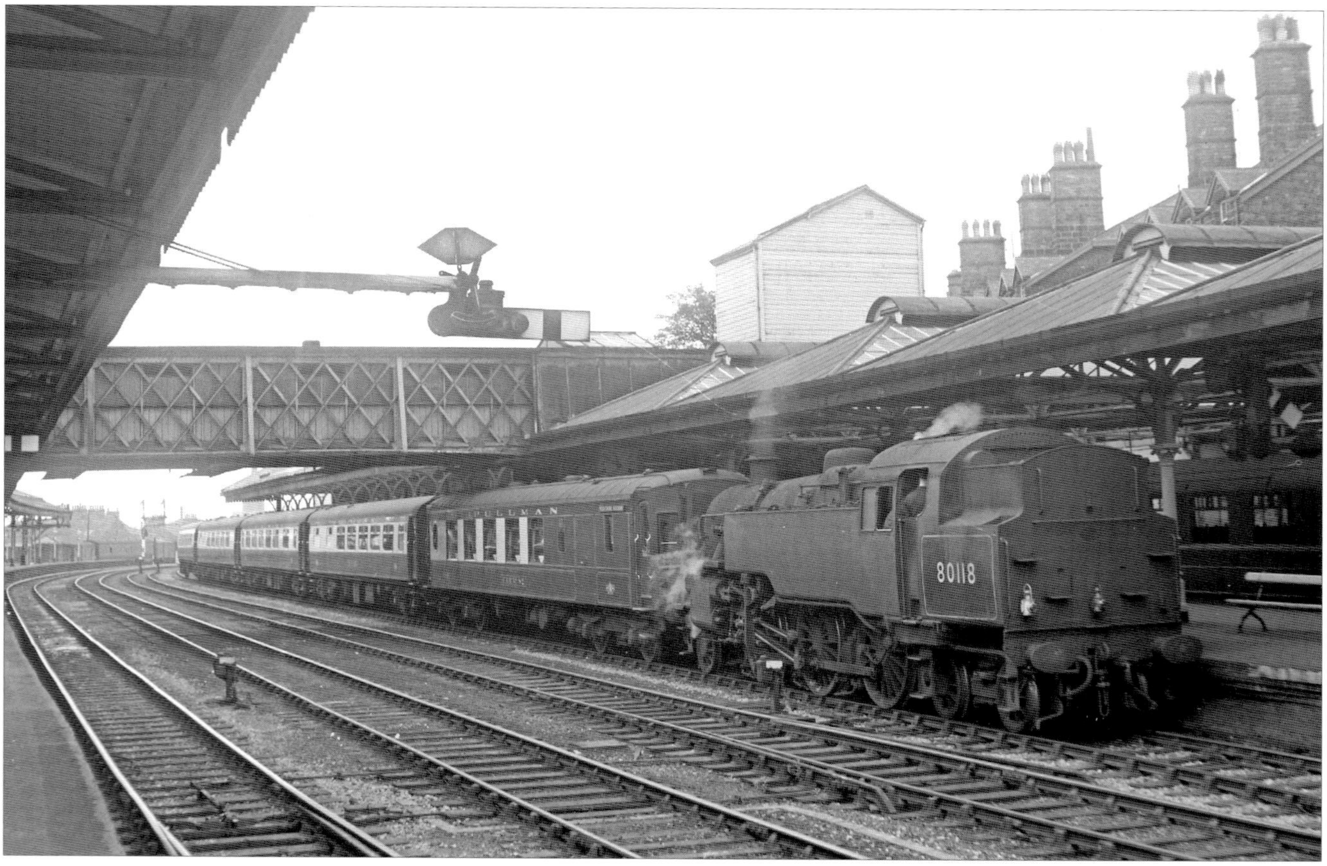
80118

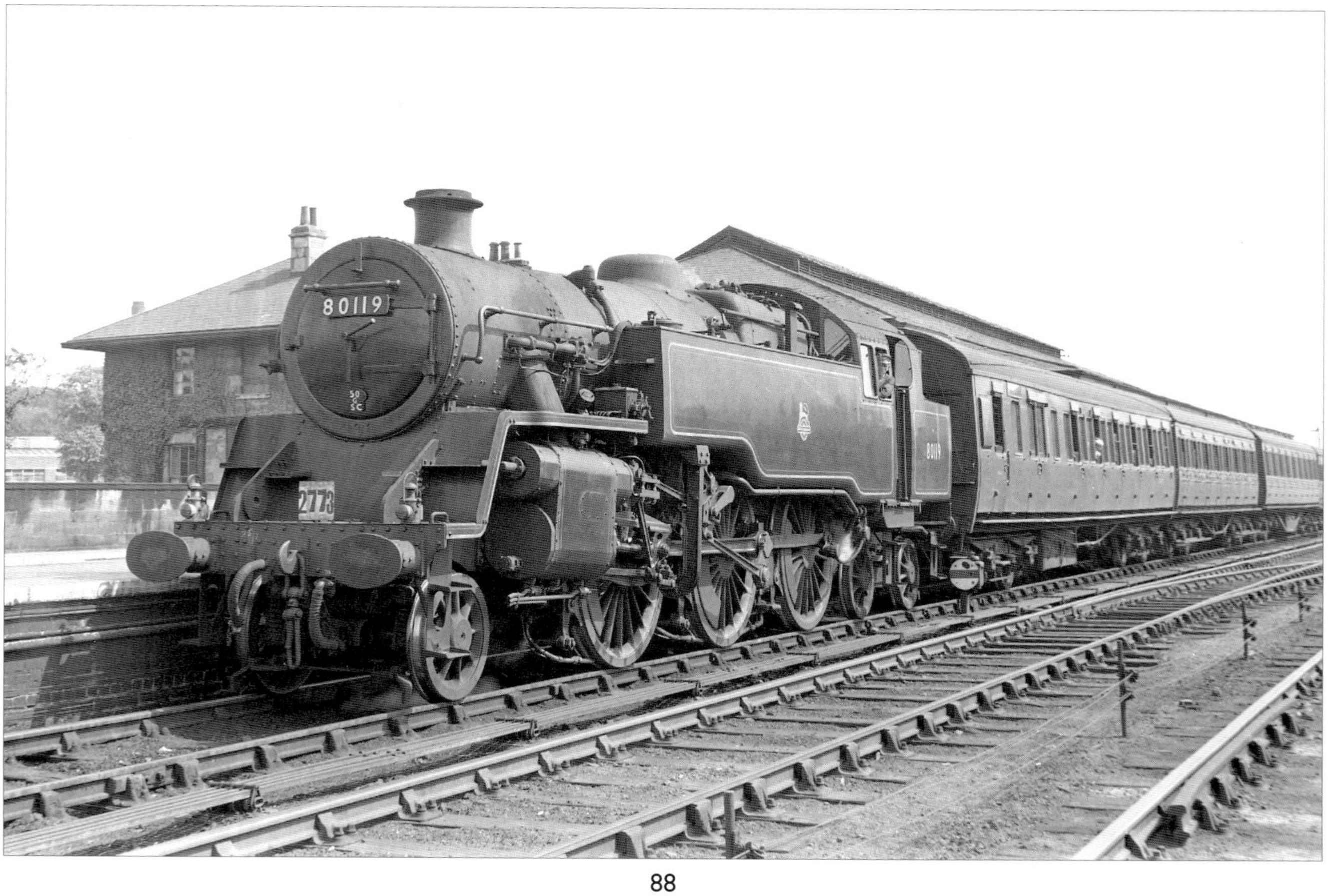
80119
2773

Top left. The Harrogate portion of the 'Yorkshire Pullman' is seen behind No 80118 on 20th April 1957, with the Class Four acting as station pilot on this occasion. The 'Yorkshire Pullman' was introduced on 30th September 1935 running from/to London King's Cross to Harrogate with a through portion to Hull via Doncaster. The service was suspended for the duration of World War 2 and resumed on 4th November 1946. The service was withdrawn by BR in 1978. No 80118 was new from Brighton Works on 15th June 1955 and allocated to Whitby where it remained until late spring 1958 when it was reallocated to Neville Hill. A change of region took place in October 1963 when it was sent to the Scottish Region's Carstairs depot. As with a large number of locomotives, a final move to Polmadie took place in spring 1964 and it was withdrawn from there on 29th November 1966. *S. Creer*

Bottom left. Seen a little over six weeks after delivery from Brighton, No 80119 is seen at Malton on 31st July 1955. Allocated to Whitby at the time, the locomotive would be transferred to Scarborough in early 1956, only to return four months later. A final move within the North Eastern Region saw it at Neville Hill before heading north along with classmates 80117/8 and 80120 in October 1963. Having arrived at Carstairs it was reallocated to Dumfries from where it was withdrawn on 29th May 1965, having been in traffic for a little under 10 years. Malton is on the line from York to Scarborough, being built by the York & North Midland Railway whose chairman was the 'railway king' George Hudson who had business interests in Scarborough (the 'Brighton of the north') and Whitby where he hoped to further develop the harbour; it opened on 7th July 1845 with the York-Scarborough line. A line from Rillington to Pickering opened on the same date with, initially, horse drawn trains running between Rillington, Pickering and Whitby. Malton had a small engine shed that opened in 1853 to supply locomotives to work local passenger and goods trains; it closed on 15th April 1963. The goods yard was closed on 3rd September 1984 with some of the last traffic being domestic coal.

Above. The 12.45pm Inverness-Aberdeen (via Craigellachie) service is seen at Elgin on the 5th September 1959. No 80121 was allocated new to the Scottish Region on 22nd July 1955 when it was allocated to Kittybrewster where it remained until that November when it was transferred to Keith. Its next move, in June 1961, was to Aberdeen's Ferryhill depot where it stayed for only a short period of time before a final transfer to Polmadie. t was withdrawn on 2nd June 1966 and consigned to Shipbreaking Industries at Faslane for breaking up. The first station in Elgin was opened by the Morayshire Railway (from 1880 the Great North of Scotland Railway (GNSR)) on 10th August 1852. The second, owned by the Inverness & Aberdeen Junction Railway (from 1865 the Highland Railway), was opened on 25th March 1858 and was later known as Elgin West. The GNSR lines to Lossiemouth and Craigellachie (where it joined the Strathspey line) were subsequently joined by the GNSR's Morayshire coastline line in 1886-7. All three of the GNSR routes were closed in the 1960s as a result of the Beeching Axe, with the Lossiemouth branch the first to go in April 1964; its station (known as Elgin East) was closed with the end of services on the coast and Craigellachie lines on 6th May 1968. The GNSR station building is still used as office accommodation and stands on the site of the original Morayshire Railway station. The present station, formerly the West (ex-Highland) station, was retained in 1968-69 and was rebuilt and the platforms were raised. *David Idle*

No 80122 is leaving Keith Junction for Aberdeen on the double-track GNSR main line east of Keith – the HR and GNSR to the west were both single track on 17th September 1955. The station was originally owned by the Highland Railway and was known as Keith Junction, the line from the west having opened by the Inverness & Aberdeen Junction Railway in 1858 and becoming part of the Highland Railway in 1865. It was the point where the line from Inverness made an end-on junction with the Great North of Scotland Railway from Aberdeen (which opened in 1856) to enable exchange of goods and passengers. As built, it was located in the vee of the routes to Inverness and to Dufftown (which diverges to the south-west here) and had four platforms – two of which were east-facing bays for local services. The Dufftown and Craigellachie line was closed to passengers by BR on 6th May 1968 as a result of the Beeching report. No 80122 was new to Kittybrewster on 8th August 1955 and was reallocated to Keith that November. It left in June 1961 for Dalry Road, staying for just under a year when it was reallocated to Greenock (Ladyburn) saw its withdrawal on 31st December 1966, with a destination of Faslane for scrapping; it was however broken up on site at Polmadie in April 1967.

New to traffic on 9th September 1955, No 80123 made the long journey from Brighton to Dundee's Tay Bridge depot. The shed was opened by the North British Railway on 1st January 1878. The original facilities included a ramped coaling stage, this was however later replaced by a mechanical plant, water tank and turntable. The shed closed on 1st May 1967 and was subsequently demolished. The only transfer of No 80123 was almost 10 years after its arrival when it was reallocated to Polmadie from where it was withdrawn on 17th August 1966, only to join numerous classmates on the scrap piles at Faslane that November. *A. W. Battson*

Top left. No 80127 was allocated new to Corkerhill depot on 7th November 1955 where it would remain for the next 8 years, eight months, and a few days, before its withdrawal on 30th July 1964. By comparison it is alongside No 40612, a Midland Railway-design, built by the LMS in December 1928, that had a service life of almost 33 years. It was withdrawn on 2nd October 1961. The locomotives are seen at Glasgow's St Enoch station on 7th September 1959. Located on St Enoch Square in the city centre, it was opened by the City of Glasgow Union Railway with the first passenger train stopped there on 1st May 1876 and the official opening took place on 17th October 1876. In 1883 it was taken over by the Glasgow & South Western Railway (G&SWR) and it became their headquarters. Services ran to most parts of the G&SWR system, including Ayr, Dumfries, Carlisle, Kilmarnock and Stranraer. In partnership with the Midland Railway, through services also ran to England, using the Settle & Carlisle Railway from Carlisle to Leeds, Sheffield, Derby and London St Pancras; the so-called Thames-Clyde route. It was a large station with 12 platforms and two impressive semi-cylindrical glass/iron roofed train sheds. The station was closed on 27th June 1966 as part of the rationalisation of the railway system following Dr Beeching's report. Upon closure its 250 trains and 23,000 passengers a day were diverted to Central. *F. W. Goudie (FWG923)*

Bottom left. Allocated new to Corkerhill on 23rd November 1955 No 80128 is seen four years late approaching Lugton station with a suburban service from Glasgow in 1959. It was transferred to Hurlford during October 1961, moving back to Corkerhill the following January. It was withdrawn on 4th April 1967 and sold to P. W. McLellan, Langloan, for disposal – one of only nine locomotives to be scrapped by the company. The station here was opened on 27th March 1871 by the Glasgow & Kilmarnock Joint Line Committee (Caledonian and G&SW Railways). The station closed to general freight traffic on 5th October 1964 and to passengers on 7th November 1966. In the loop beside the train is AC Cars railbus No Sc79979 which will form a connecting service from Lugton to Beith. The Beith branch closed on 5th November 1962. *N. Stead Collection (NS207845)*

Above. Sitting outside Brighton Works on 8th December 1955 is No 80129, it was officially released to traffic the following day and would make the long journey to Polmadie. Of note for this batch, Nos 80121-130, is the inclusion of a recess for a tablet catcher in the cab side and fitting of a speedometer. The locomotive was withdrawn week ending 7th July 1964 and subsequently reinstated and allocated to Lostock Hall, from where it would finally be withdrawn on 10th October the same year. It would be cut up at Crewe Works the following month. Brighton railway works (also known as Brighton locomotive works, or just the Brighton works) was one of the earliest railway-owned locomotive repair works. Founded in 1840 by the London & Brighton Railway and thus pre-dating the more famous railway works at Crewe, Doncaster and Swindon. The works grew steadily between 1841 and 1900 but efficient operation was always hampered by the restricted site, and there were several plans to close it and move the facility elsewhere. Nevertheless, between 1852 and 1957 more than 1,200 steam locomotives as well as prototype diesel electric and electric locomotives were constructed there, before the eventual closure of the facility in 1962, and demolition seven years later. The motive power depot, next to the works, was officially closed on 15th June 1961, but remained in use for stabling steam locomotives until 1964, and was demolished in 1966.

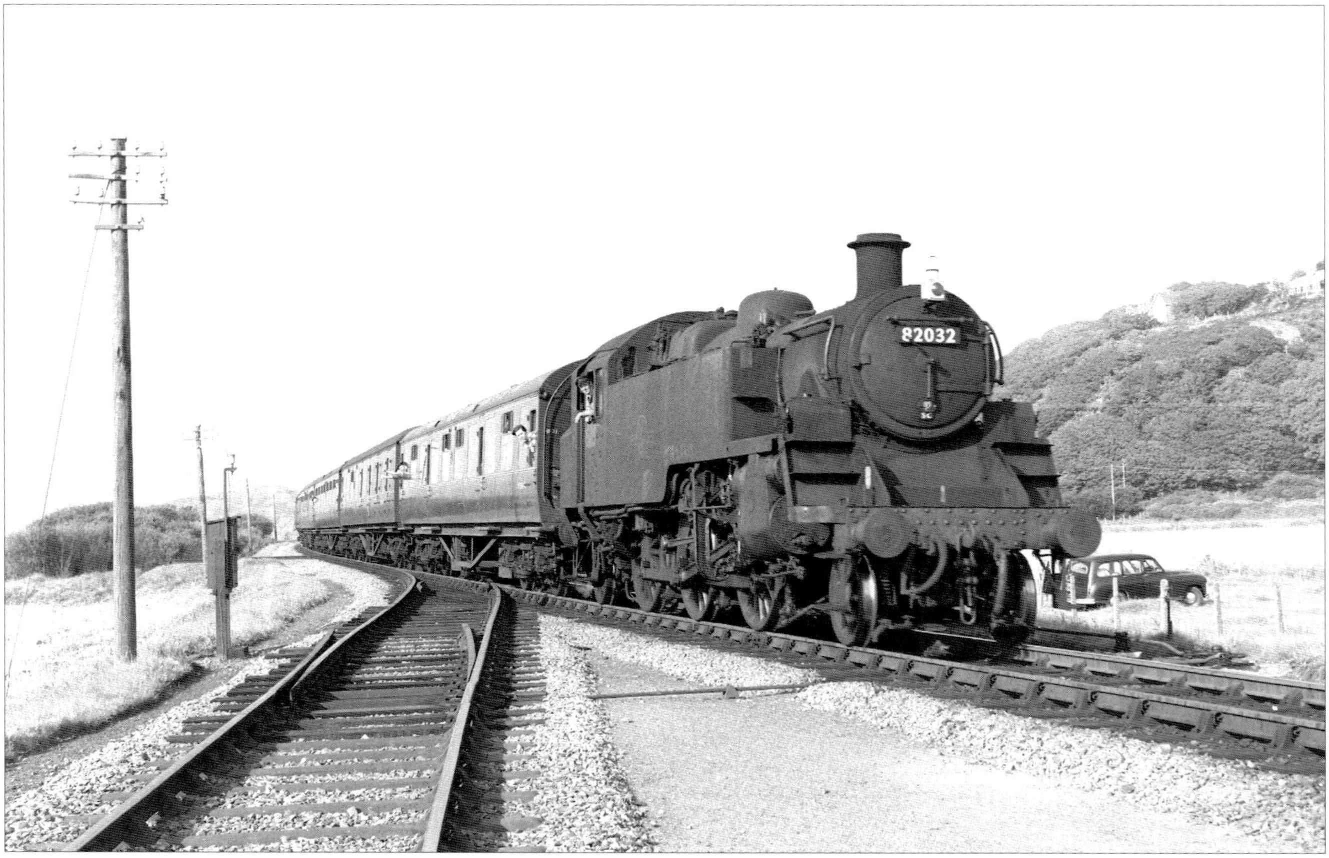
82032

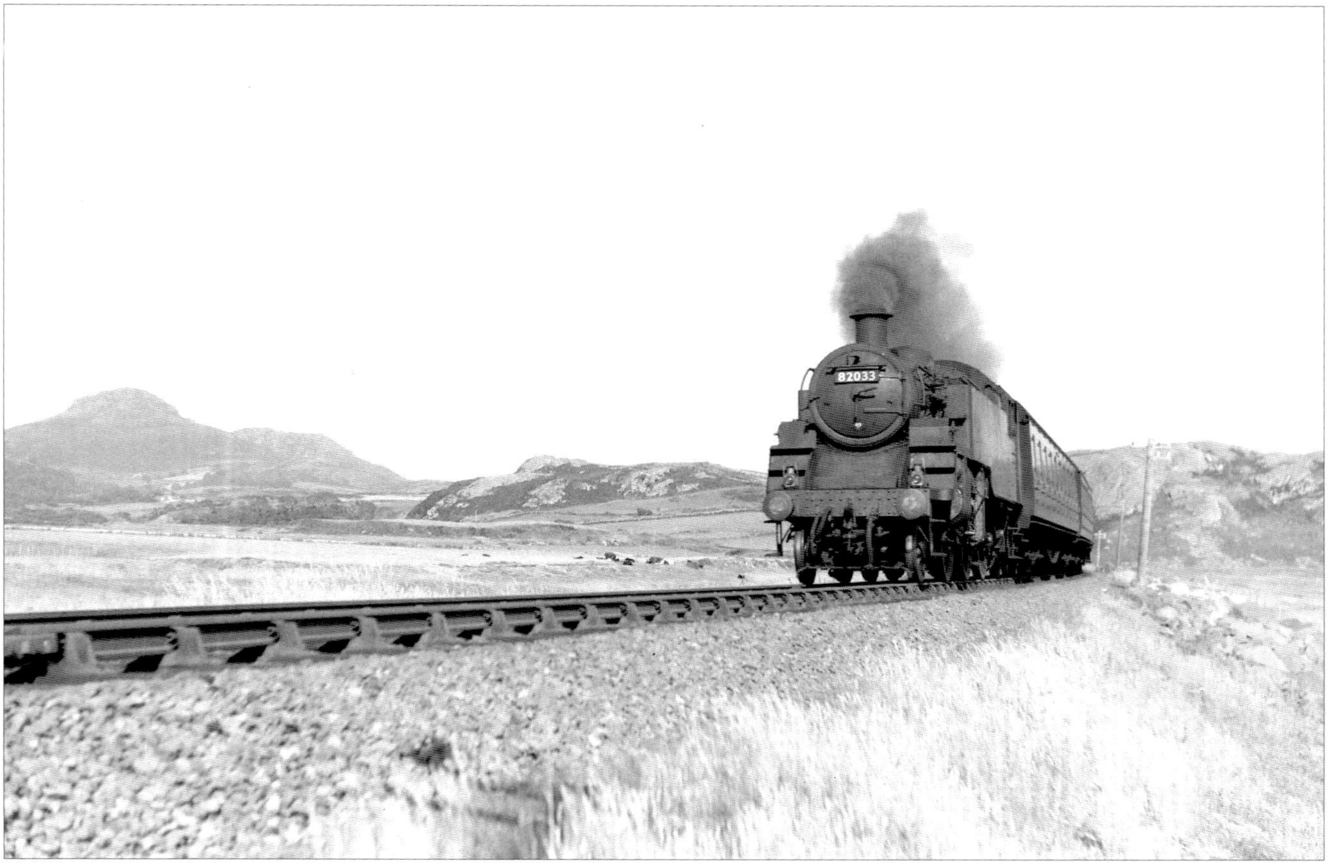
82033

Top left. No 82032 is seen approaching Fairbourne with the 4.05pm Pwllheli-Machynlleth service on 30th July 1963. New from Swindon Works on 1st January 1955, it was allocated to Barry. As with most of the class it was to move around a fair bit with allocations to: Newton Abbot, Treherbert, Radyr, Bristol Bath Road and both Chester depots before arriving at Shrewsbury on 23rd April 1961. A few months later it was reallocated to Machynlleth before a return to Shrewsbury by the end of the year. It was transferred to Bangor in June 1964, from where it was withdrawn on 1st May 1965. The station was opened as Barmouth Ferry by the Aberystwyth & Welsh Coast Railway on 3rd July 1865, the Cambrian Railways absorbed the company two days later. The station was to close less than two years later. It was reopened as Fairbourne on 6th June 1899. *B. Wadey*

Bottom left. What the photographer recorded as the Down 'Cambrian Coast Express' is seen at Criccieth on 1st July 1961 behind No 82033. New to Newton Abbot on 13th January 1955, the locomotive arrived for use over the Cambrian line in December 1960 when it arrived at Machynlleth. A move to Bangor took place in June 1964 with a final move south to London's Nine Elms depot where it arrived in late April 1965. It would survive for another five months, being withdrawn on 19th September the same year. The first official use of the name 'Cambrian Coast Express' was in 1927 when the train ran only on summer Fridays and Saturdays. By 1939 the through train was running from Paddington to the Welsh coast only on summer Saturdays. After World War 2, the service was re-introduced on Saturdays only and its seasonal operation continued under BR, usually with through coaches to both Aberystwyth and Pwllheli. By 1957 it was running every day except Sundays all year round. A change of motive power took place at Shrewsbury as most main line locomotives were too heavy for use over the Cambrian line. *J. Flint/J. Harbart (FH941)*

Above. The signalman and fireman are about to exchange tokens as No 82034 arrives at Morfa Mawddach heading for Machynlleth. The station was built by the Aberystwyth & Welsh Coast Railway, opening on 3rd July 1865 as Barmouth Junction. From 1899 to 1903 there was a connection with the Barmouth Junction & Arthog Tramway. The station was host to a GWR camp coach from 1934 to 1939. A camping coach was also positioned here by the Western Region from 1956 to 1962. In 1963 the administration of camping coaches at the station was taken over by the London Midland Region, there were three coaches here in 1963 and 1964 and two from 1965 to 1968. Until the 1960s there was a summer service between London Paddington and Pwllheli, via Birmingham Snow Hill, Shrewsbury and Machynlleth. On 13th June 1960 it was renamed Morfa Mawddach. No 82034 was new to Newton Abbot on 28th January 1955, with subsequent transfers to Wellington, Treherbert, Radyr, Bristol Bath Road and Chester before being allocated to Machynlleth on 23rd April 1961. A transfer to Patricroft in spring 1965 saw its withdrawal on 10th December 1966. *J. Flint/J. Harbart (FH1917)*

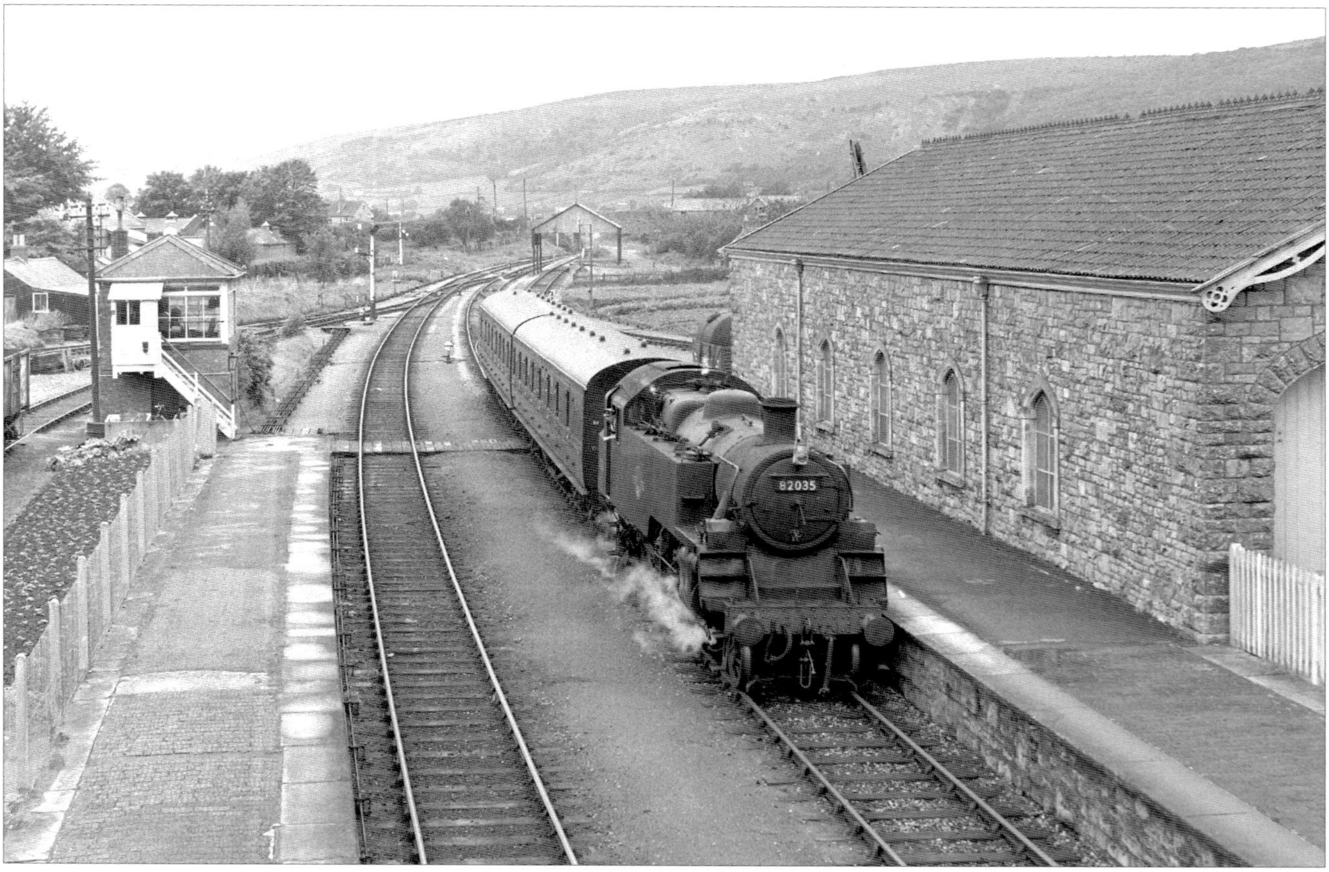
82035

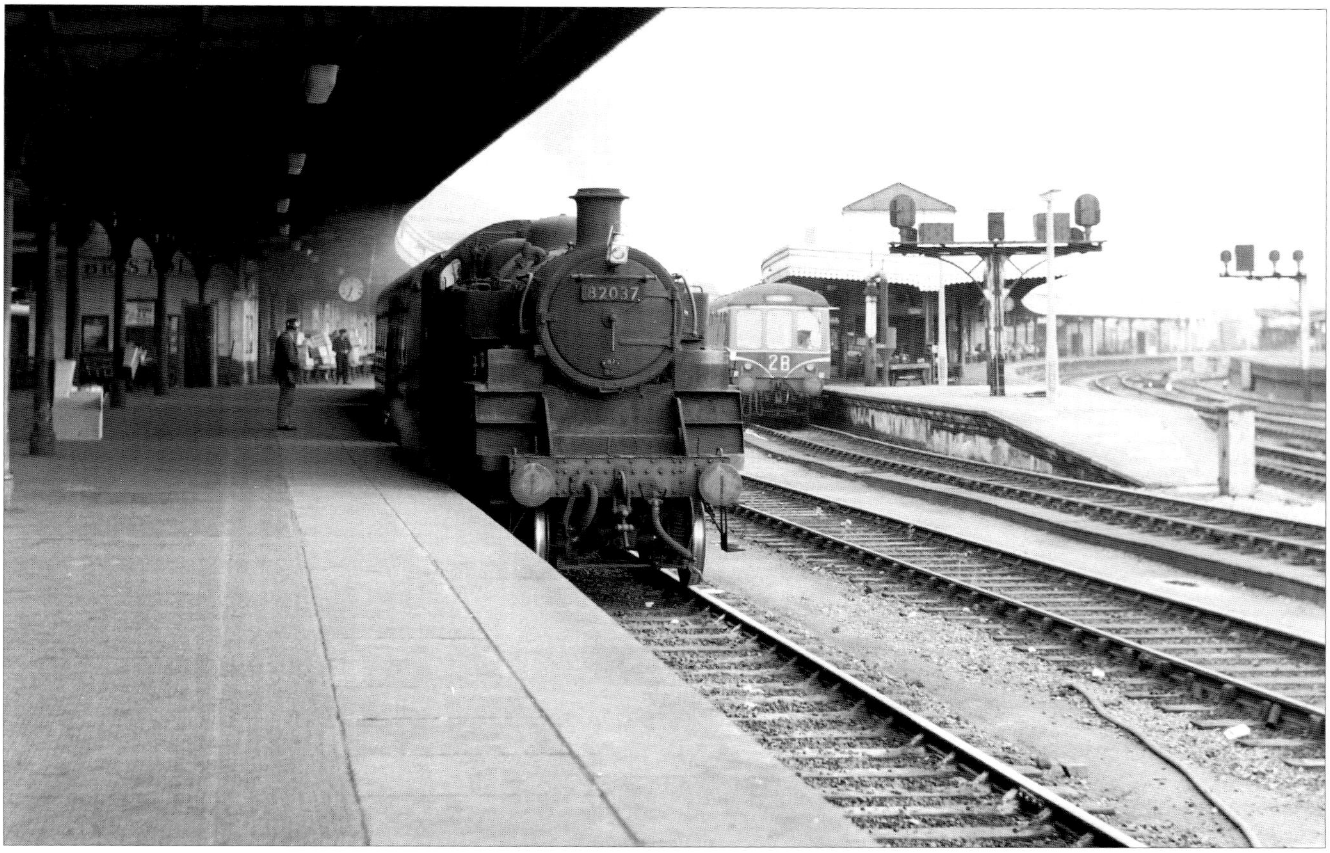
82037
2B

Top left. The 2.45pm Yatton-Witham service is seen at Cheddar on 15th September 1962. The station opened as the temporary terminus of the Bristol & Exeter Railway's broad gauge line in August 1869. The railway was extended to Wells in 1870, converted to standard gauge in the mid-1870s and then linked up to the East Somerset Railway to provide through services from Yatton to Witham in 1878. All the railways involved were absorbed into the GWR in the 1870s. The Yatton to Witham line closed to passengers on 9th September 1963. Cheddar remained open for goods until 29th November 1965, and even then a private siding kept the line in place until March 1969. No 82035 was new from Swindon Works on 3rd March 1955, being allocated to Barry where it remained until the spring of 1958 when it was reallocated to Bristol – spending time at both St Philip's Marsh and Barrow Road It moved to the former Southern Region depot at Exmouth Junction in spring 1964. In late summer 1964 it moved to Yeovil from where it was withdrawn on 6th August 1965. *B. Wadey*

Bottom left. Bristol Temple Meads on 15th July 1963 with No 82037 at the head of a service for Portishead. The original terminus was built in 1839-41 for the GWR, the first passenger railway in Bristol, and was designed by Isambard Kingdom Brunel, the railway's engineer. It was built to accommodate Brunel's 7ft 0¼in (2,140mm) broad gauge. The station was on a viaduct to raise it above the level of the Floating Harbour and River Avon, the latter being crossed via the Grade I listed Avon Bridge. The station was also used by the Bristol & Exeter Railway, the Bristol & Gloucester Railway, the Bristol Harbour Railway and the Bristol & South Wales Union Railway. Its name 'Temple Meads' derives from the nearby Temple Church, which was gutted by bombing during World War 2. Train services to Bath commenced on 31st August 1840 and were extended to Paddington on 30th June 1841 following the completion of Box Tunnel. No 82037 was released to traffic on 20th April 1955 and allocated to Swansea Paxton Street depot that served the town's Llanelly Railway & Dock Co's Victoria station. It was then moved around the Western Region serving depots in Wales and England before ending its career at Bristol Barrow Road on 25th August 1965; making the final journey to Newport for dismantling by Cashmore's cutters. *A. E. Bennett (B6332)*

Above. The goods area of Exeter Central station is not often photographed but it was where we find No 82039 on 1st January 1965. New from Swindon Works on 10th May 1955, the locomotive was allocated to Barry where it remained until transfer to Bristol Bath Road in the spring of 1958. Christmas 1958 saw its transfer to the former S&D depot at Templecombe before returning to Bristol's Barrow Road shed in late summer 1960. Eighteen months later it moved across the City to St Philip's Marsh, before moving to Exmouth Junction via Barrow Road in spring 1964. A final relocation to Gloucester Horton Road in late spring 1965 saw its withdrawal on 2nd July 1965. The LSWR opened its Exeter extension from Yeovil Junction on 19th July 1860 and its station at Queen Street in the city centre became the terminus for services from London Waterloo station, known as Exeter Queen Street. From 1st May 1861 it was also the terminus for trains on the new Exeter & Exmouth Railway. This was also operated by the LSWR but the physical junction between the two lines was at Exmouth Junction, 1.1 miles (1.8km) east of Queen Street. The final piece of the LSWR's network in Exeter was opened on 1st February 1862 when a steep line descended from the west end of Queen Street station to reach the Bristol & Exeter Railway's station at Exeter St Davids which had been opened in 1844. Here the LSWR connected with the Exeter & Crediton Railway and, over that line, eventually reached Plymouth, Padstow, Bude, and Ilfracombe. The station was renamed Exeter Central on 1st July 1933 on completion of a rebuilding programme.

No 82043 takes its three coach train over Station Road level crossing at Pilning on the Avonmouth branch with the signalman and fireman about to exchange tokens. Opened as 'Pilning' in 1863 by the Bristol & South Wales Union Railway, later part of the Great Western Railway, on the line from Bristol to New Passage Pier (for ferry connections across the River Severn to Portskewett). This station closed on 1st December 1886 at the same time as the Severn railway tunnel opened. However, it reopened (as 'Pilning Low Level', to differentiate it from the main line station) in 1928 when a new rail connection was built to Severn Beach. The station finally closed on 23rd November 1964, with the line closing in 1968 – all freight services were concentrated at the main line station. No 82043 was new to Barry on 27th June 1955, remaining there until late spring 1958 when it was transferred to Bristol Bath Road. It remained a local engine with allocations at both St Philip's Marsh and Barrow Road from where it was withdrawn on 7th February 1964. It was destined to be scrapped at Eastleigh Works during the first week of May.

The final member of the class was ex-works on 12th August 1955 and allocated to Barry, joining No 84043 on the trip to Bristol. After stints at both Bath Road and Barrow Road depots it made the trip to Neyland in late summer 1961; it returned east to Taunton depot within a matter of weeks. Late spring 1964 saw it allocated to Exmouth Junction before a move to Gloucester Horton Road a year later. It was withdrawn on 2nd August 1965, only to be reinstated on 20th September when it was allocated to Templecombe's S&D depot. It was finally condemned on 18th November 1965, returning to South Wales for disposal. It is seen here approaching Yeovil Pen Mill with a service from Taunton on 25th August 1962. The station was opened by the GWR as part of the Wilts, Somerset & Weymouth route on 1st September 1856. The route was completed to Weymouth on 20th January 1857. The Bristol & Exeter Railway's line from Taunton, which initially terminated at Yeovil Hendford, was extended to connect with the GWR at Yeovil Pen Mill from 2nd February 1857. Both these lines were built using the 7ft 0$\frac{1}{4}$(2,134mm) broad gauge. *B. Wadey*

# The Class of 1956

The Oxford University Railway Society's rail tour in March 1964 appears to have gone unrecorded! It is seen here taking water at Moat Lane Junction behind No 80131 on the line between Llanymynech and Aberystwyth. The first station at Moat Lane was opened in 1859 by the Newtown & Machynlleth Railway, being located a short distance to the south-west of the later junction station, and was intended to serve Caersws. On the opening of the Llanidloes & Newtown Railway's Machynlleth line in 1863 a new station was built in the 'V' of the junction and replaced the original station. This had a single straight platform face serving Llanidloes trains and a curved platform serving the new line. Since the latter was now the major route, an island platform was also provided on the curve, providing three platforms for trains to/from the Machynlleth direction. The station closed on 31st December 1962, together with passenger services to Llanidloes and Brecon. No 80131 was new to Plaistow on 2nd March 1956, remaining on the LT&SR until moving west to Old Oak Common on 15th July 1962. That September saw it move to Shrewsbury, and on to Oswestry in January 1963. Two years later it was allocated to Bangor from where it was withdrawn on 8th May 1965. *J. Pearse*

The 9.48am Gloucester to Hereford service is seen at Mitcheldean Road behind No 80135 on 31st October 1964. The station was opened in 1855 by the Hereford, Ross & Gloucester Railway, linking Ross-on-Wye to Grange Court where it met the Gloucester & Dean Forest Railway. Both companies were taken over by the GWR on 29th July 1862. Goods services at the station ceased on 12th August 1963, followed by passenger traffic on 2nd November 1964 – the line closed throughout the following year. No 80135 was new to traffic from Brighton Works on 30th April 1956, being sent to the LT&SR's Plaistow depot. Following a transfer to Tilbury in November 1959 it was reallocated to the Western Region's Shrewsbury shed on 15th July 1962. A move to Oswestry the following spring was followed by a return to Shrewsbury in September 1964, from where it was withdrawn on 24th July 1965. Sold to Woodham Bros, at Barry, it was subsequently saved for preservation.

Above. Photographed from a foot crossing at an unidentified location within the third-rail network is No 80139. The image is dated as 1956, with the head code, amongst others, for 'Brighton and Salisbury via Southampton Central', it may well be on a running in turn before heading to the LT&SR. No 80139 was new to Neasden on 26th June 1956 and stayed until reallocated to the Southern Region in December 1959 when it arrived at Tunbridge Wells West. It stayed there until late summer 1963 when it was transferred to Brighton, moving onto Redhill that Christmas. A final move took it to Eastleigh in early spring 1965; it survived until the end of steam on the Southern on 9th July 1967.

Top right. No 80140 is seen here light engine taking water at Southampton Central on 5th September 1966. New to Neasden on 10th July 1956, it departed for the Southern Region's Tunbridge Wells West depot in December 1959. Following allocations at Brighton and Redhill it arrived at Feltham depot in early summer 1965 – its home shed at the time the photograph was taken. A final move to Eastleigh took place in late summer 1966; it was to survive until the end of steam on the Southern on 9th July 1967.

Bottom right. The sight of double-headed Class Fours are not that common, however, on 3rd October 1963 Nos 80141 and 80088 are heading a Tunbridge Wells West-Eastbourne service south of Groombridge junction. No 80141 was new to Neasden on 25th July 1956 and as with most of this batch headed south of the Thames. It probably travelled with No 80140 on its journey towards oblivion. Withdrawn on 9th January 1966 it met it fate at the hands of Cox & Danks cutters at Park Royal two months later. The story of No 80088 is recounted on page 67. *S. Creer*

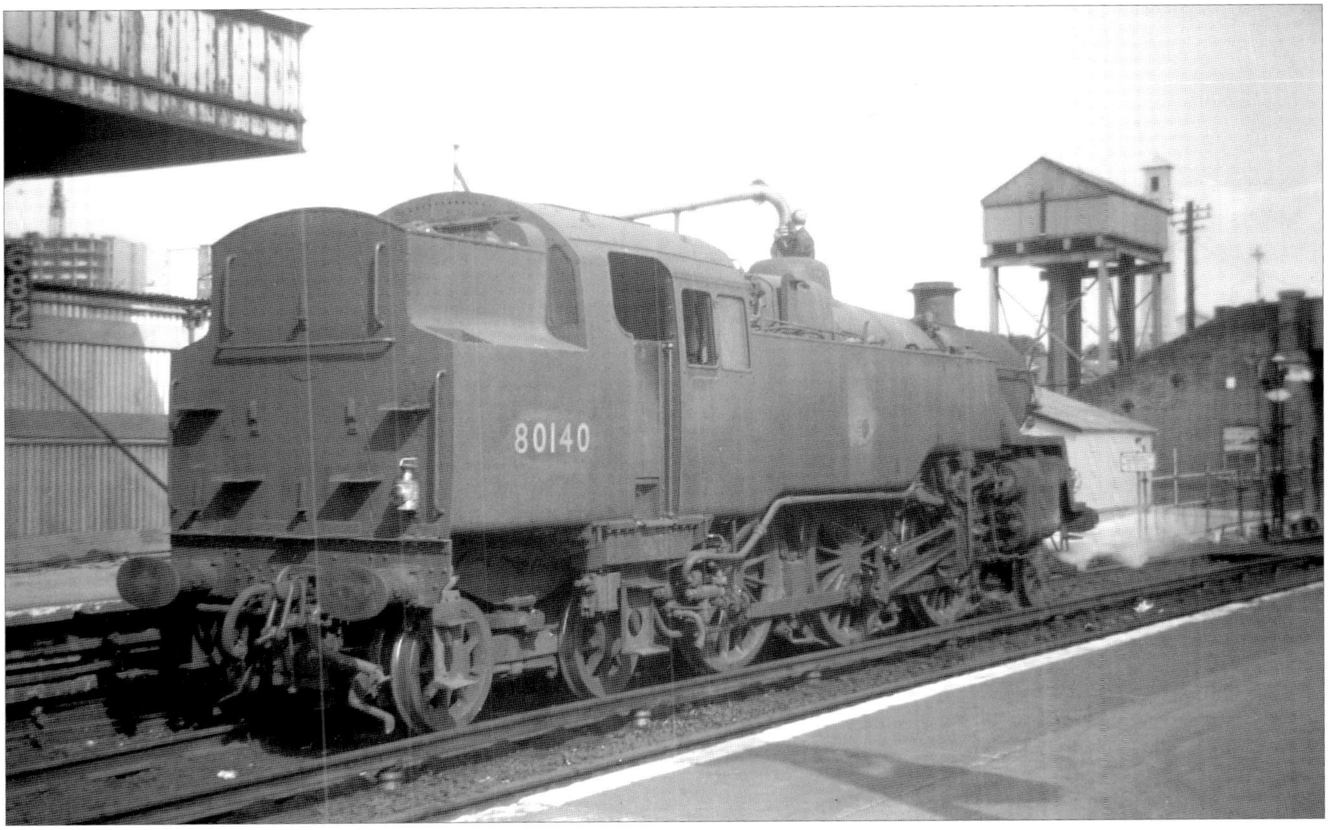
80140

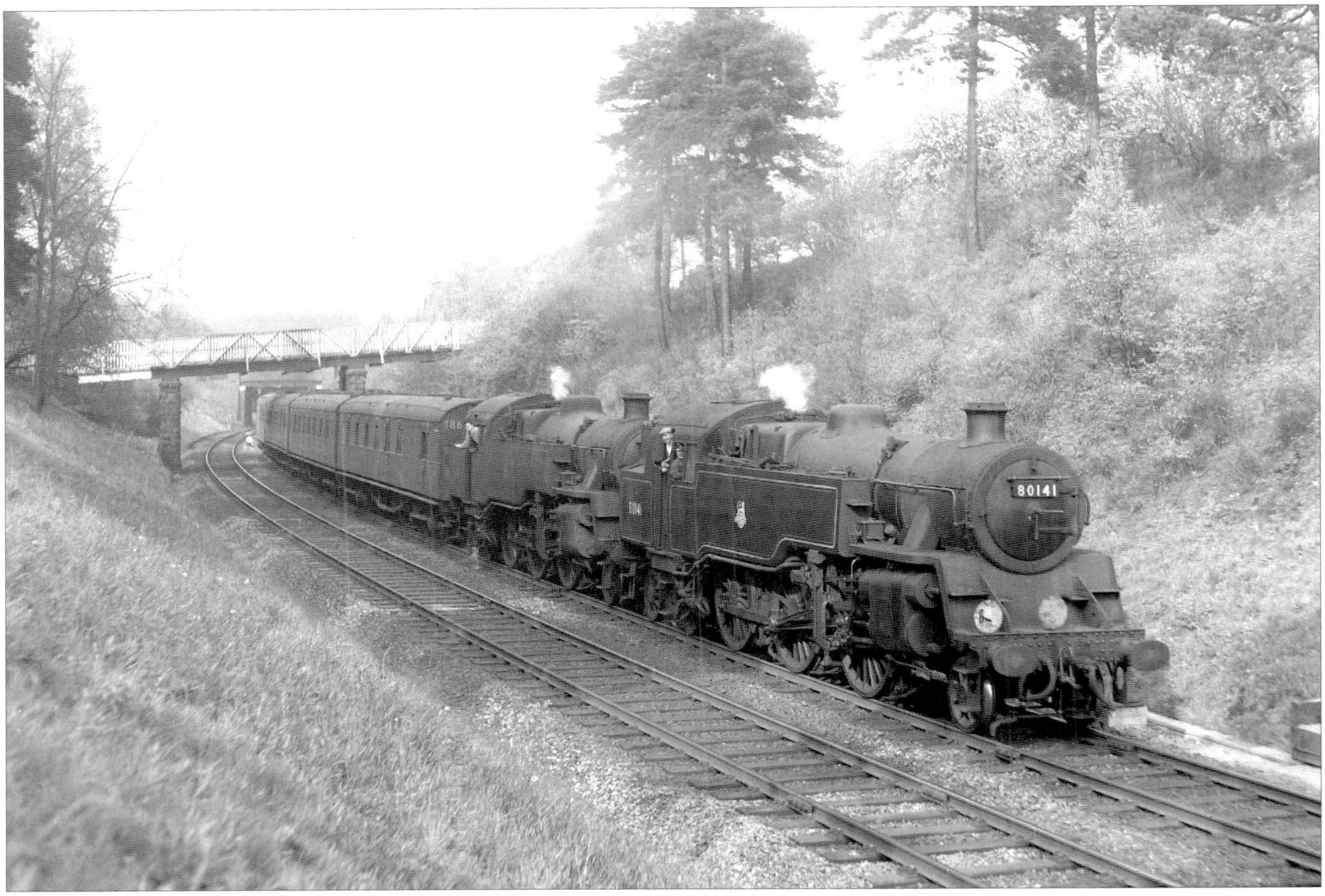
80141

# British Railways Standard Tanks

Below. No 80145 and its train are seen at West Hoathly's platform as the passengers alight, probably soon after delivery as it has yet to receive a shed plate. The station is on what is now the Bluebell Railway and despite its name the site of the station is actually situated in the village of Sharpthorne, half a mile from the village West Hoathly. The station opened in 1882, just north of the 731yds (668m) long Sharpthorne Tunnel. Along with other stations along the line, all constructed under the influence of the LBSCR, a then substantial provision of £17,000 was made to construct each two-platform through station (the architecture was similar to that now seen at the restored Kingscote). The station was finally closed on 17th March 1958, but was used by contractors demolishing the line in the 1960s to bring equipment in and out. No 80145 was new to traffic from Brighton Works on 16th October 1956 and moved next door to the depot where it remained until December 1963 when it was reallocated to Redhill. The summer of 1965 saw it allocated to Salisbury where it remained until early 1966 when it made its final move to Nine Elms. It almost survived until the end of Southern steam being withdrawn on 25th June 1967, by the end of the year it was scattered around Cashmore's Newport yard where it was dismantled. *R. K. Collins*

Top right. The crew of No 80149 look towards the cameraman as it stands in the platform at Horsted Keynes with a rake of condemned wagons. The station was built in 1882 by the LBSCR, and is today one of the operational intermediate stations on the Bluebell Railway. It was a junction station, with a line branching off to Haywards Heath via Ardingly that was opened by the LBSCR on 3rd September 1888 – the branch was electrified from 7th July 1935, closing to freight on 2nd April 1962 and to passengers on 28th October 1963. The East Grinstead-Lewes line was initially closed to passenger services on 28th May 1955, however a legal challenge – based on the Act of Parliament to construct the line, saw the stations specified in the it reopened on 7th August 1956. Not to be beaten BR finally closed the line on 17th March 1958. Freight services had already ceased on 5th March 1962. No 80149 was new from Brighton Works on 13th December 1956, and allocated to the adjacent shed. Its one and only relocation took place around Christmas 1963 when it moved to Redhill. It was withdrawn on 7th March 1965, and was dismantled at Park Royal by Cox & Danks in June.

Bottom right. Wearing its 75A Brighton shed plate, No 80150 stands in the works yard at Eastleigh on 22nd May 1960. New from Brighton Works on 28th December 1956, it was transferred to Eastleigh depot in late summer 1963. Withdrawn on 17th October 1965, it was destined to be scrapped in Woodham's yard at Barry and was one of those that survived to be preserved. *David Idle*

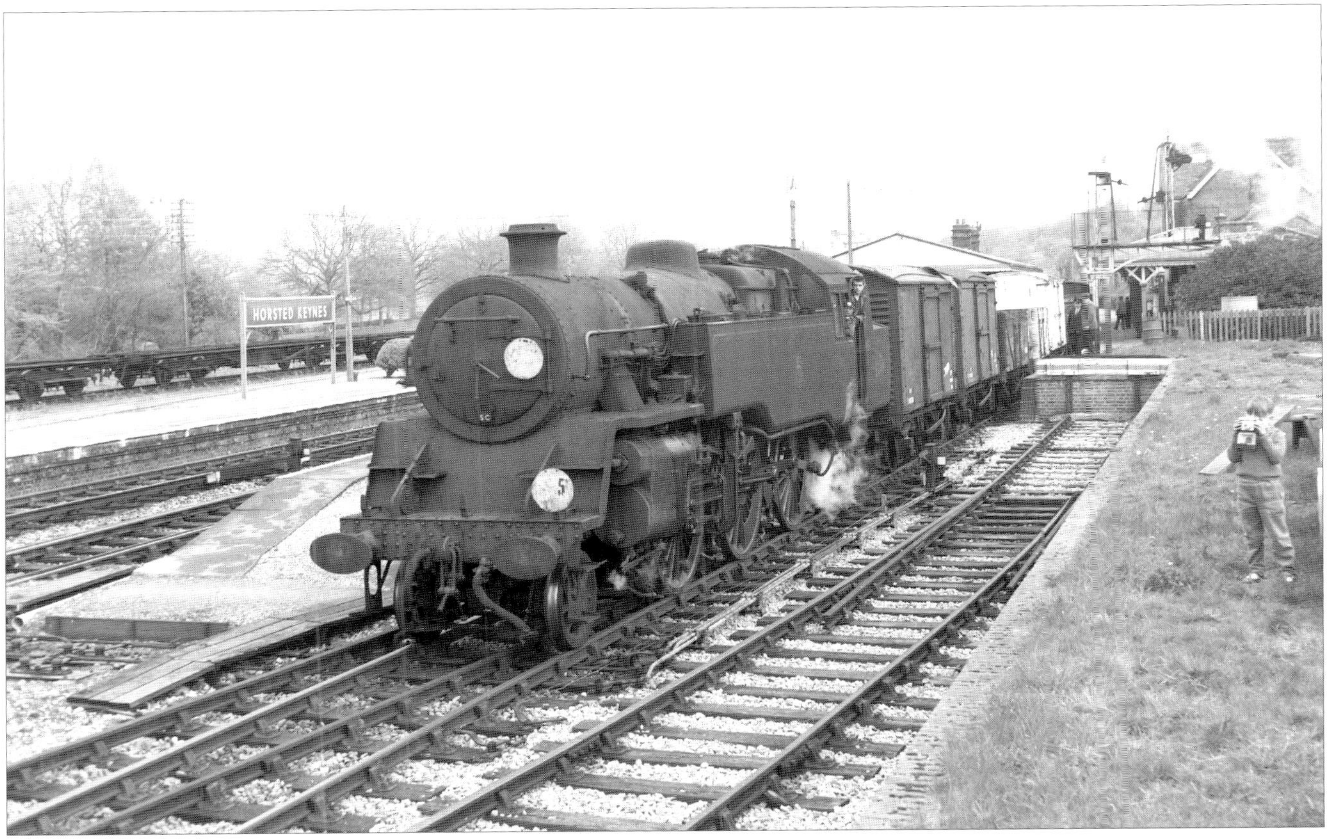
HORSTED KEYNES

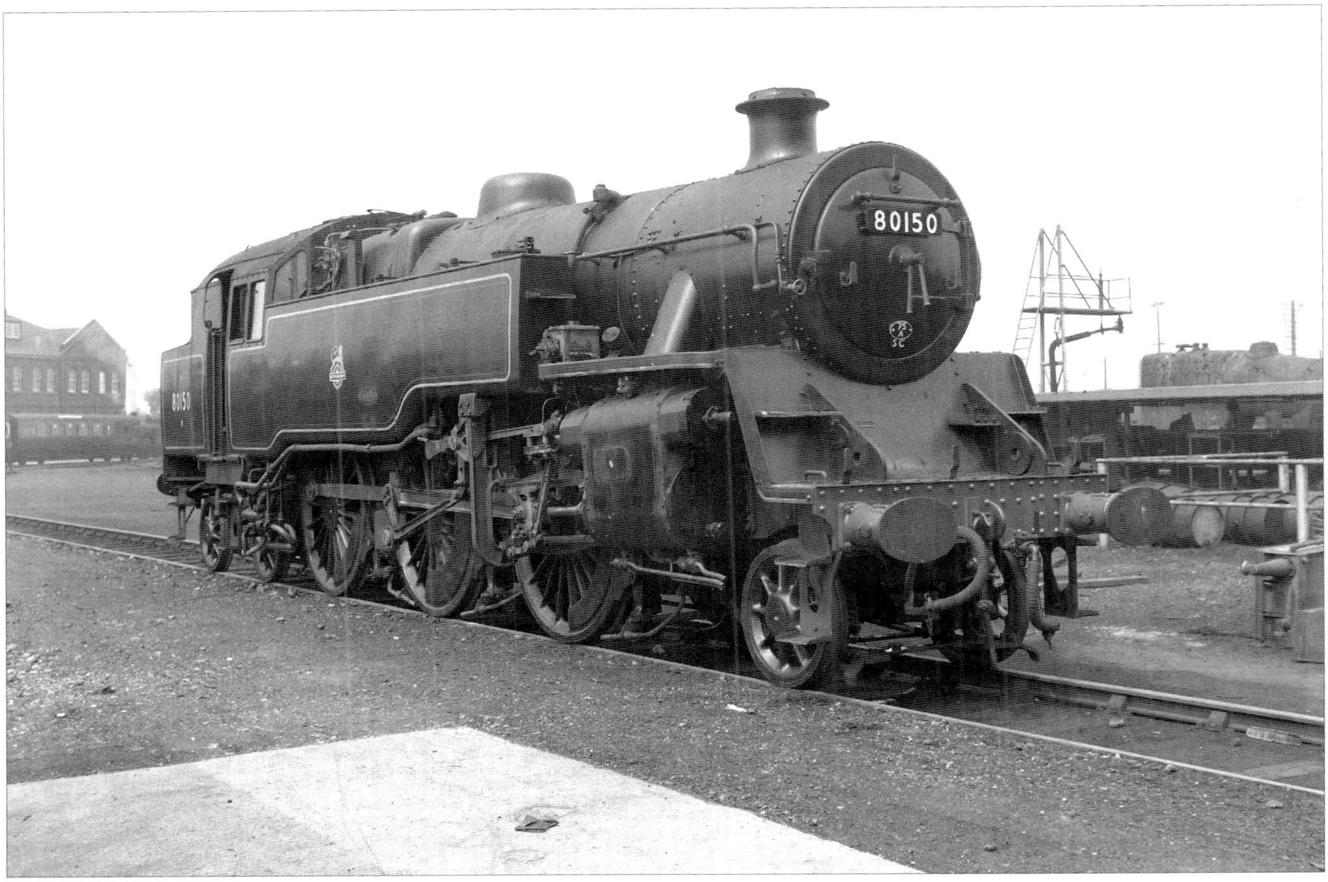
80150

The Class of 1957

# The Class of 1957

Top left. Steam on the Southern has only a few months to go as No 80151 stands in the shed yard at Eastleigh carrying a Lymington destination board and a 'home made' number plate on the smoke box door. In 1967, the Brockenhurst-Lymington Pier branch line was the last steam-operated line on the BR system. The last passenger train ran on Sunday 2nd April 1967 behind LMS Ivatt Class 2 2-6-2T No 41312; ordinarily this train of the day terminated at Lymington Town and berthed there overnight. On the final run the locomotive ran round its train at Lymington Town and the train returned empty to Brockenhurst – this was the last ever run-round of coaches by a steam engine on a UK branch line in regular service. No 80151 was new to traffic on 18th January 1957 and allocated to the shed next door. Christmas 1964 was it move to Redhill, then onto Salisbury the following summer. A final transfer in late summer 1966 saw it arrive at Eastleigh – it did not survive until the end of Southern steam being withdrawn on 7th May 1967, courtesy of Woodham's at Barry it was to survive into preservation. *E. Sawford (ES6050)*

Bottom left. Doing its best to hide in the murk at Sheffield Park is No 80152, with what appears to be a single ex-LBSCR carriage. No doubt this was taken during the period when BR operated its' sulky service' over what is now the Bluebell Railway. The station was divided into two levels: the higher-level platforms serving the Three Bridges-Tunbridge Wells Central line, whilst the lower-level platforms received services from the Oxted line and the East Grinstead-Lewes line. The Lewes & East Grinstead Railway had arrived from the south on 1st August 1882, followed by the Croydon, Oxted & East Grinstead Railway from the north on 10th March 1884. Only the lower-level platforms remain open today, the high level having closed in 1967 with the Three Bridges to Ashurst Junction line as part of the closure programme proposed by the Beeching Report. No 80152 was ex-works on 6th February 1957, and spent the following seven years at the local depot. At Christmas 1963 it was allocated to Redhill, moving to Salisbury the following summer. A final move to Eastleigh took place in late summer 1966 was followed by withdrawal on 9th July 1967 and scrapping by Birds, Risca, that November. *B. Wilson*

Below. The last of the line, No 80154, is seen at the head of a rake of parcels vans at Waterloo on 18th September 1965. New to Brighton depot on 26th March 1957 the locomotive was moved to Feltham in July 1963, moving onto Nine Elms in November 1964. It was withdrawn on 2nd April 1967, meeting its fate in South Wales at the hands of Buttigiegs cutters four months later. Waterloo was built by the LSWR, opening on 11th July 1848 as Waterloo Bridge. It was not designed to be a terminus, but a stop on an extension towards the City and was originally laid out as a through station. The LSWR purchased several properties along the route, before plans were cancelled following the 'Panic of 1847'. In October 1882, the station was officially renamed Waterloo, reflecting long-standing common usage, even in some LSWR timetables. With Waterloo now destined to remain a terminus station, and with the old station becoming a source of increasingly bad will and publicity amongst the travelling public, the LSWR decided on total rebuilding, in a project they called the 'Great Transformation'. The rebuilt station was formally opened on 21st March 1922 by Queen Mary. The main pedestrian entrance, the Victory Arch (known as Exit 5), is a memorial to company staff that were killed during World War 1. Upon opening, it marked 585 employees who had been killed in World War 1. *David Idle (DI351-33)*

Above. New to traffic from Darlington Works on 27th March 1957 No 84020 was sent south to Ashford depot. It is seen approaching Appledore whilst working an Ashford-New Romney working on 21st June 1959 – the platforms here were staggered. The station was opened on 13th February 1851 by the South Eastern Railway it became a junction station in 1881 when a branch line opened to Lydd and New Romney; this closed to passengers on 6th March 1967 following the Beeching Report, general freight services had ceased on 18th April 1964. Part of the line remained operational to service Dungeness Nuclear Power Station. In early 1961 No 84020 was transferred to Stewarts Lane, before a move westwards to Exmouth Junction that June. In September 1961 it was sent to Llandudno Junction from where it headed to Crewe as one of the works shunters. It was condemned on 31st October 1964 and scrapped by Hughes Bolckow at North Blyth. *B. Wadey*

Top right. No 84021 is seen departing Ramsgate for Ashford on 28th March 1959. To the left of the image is part of Ramsgate carriage sidings; from around 1934, four of these tracks were covered for 655ft (199.6m) of their length by a pitched-roof steel-framed structure clad with corrugated asbestos, this becoming the carriage cleaning shed. The number on the head code disc on the top bracket is the duty number. In the winter of 1957 this duty commenced at Ramsgate shed at 2.20am with the locomotive leaving the depot, finishing at 9.12pm. No 84021 was allocated to Ashford depot on 30th March 1957 following completion at Darlington Works. In early 1961 it was reallocated to Bricklayers Arms, moving west to Exmouth Junction in May the same year. Quickly followed by a change of regions, arriving at Llandudno Junction in mid-September 1961. A move to Crewe Works in July 1962 saw it acting as one of the works shunters. It was withdrawn on 5th September 1964, being cut up at the works the following month.
*R. C. Riley (13071)*

Bottom right. Duty 362 was the second half of duty 361. The day started at Ashford depot at 6.05am, finishing at Ramsgate loco in the early hours of the following morning. As part of the duty it was used as a carriage shunter at Margate between 10am and midday and, following a round trip to Ramsgate, shunting the yard again between 3pm and 9.15pm. No 84021 was new to Ashford on 30th March 1961, moving onto Bricklayers Arms and Exmouth Junction before arriving at Lancaster Green Ayre in September 1961. It was allocated to Widnes in January 1962 before joining No 84021 at Crewe Works, and sharing the same fate.

Above. Ex-Darlington Works on 10th April 1957, No 84024 was delivered new to Ashford where it served for almost four years. It is seen in the shed yard carrying duty number 363 on the head code disc. In spring 1961 it was reallocated to Bricklayers Arms, followed by a move back down south to Brighton that May. A change of region came that September when it was reallocated to Warrington Dallam before a move to Widnes in January 1962. In July it became one of the class to be used as works shunters at Crewe, joining its fellows in being withdrawn on 5th September 1964. The first depot at Ashford was opened by the Maidstone & Ashford Railway in 1894; enjoying only a short existence, it was replaced in 1899 by the South Eastern Railway. In 1931 the Southern Railway opened a new 10-road shed that served until the end of Southern steam in the area in June 1962.

Top right. With the unmistakable feature of Shakespeare Cliff in the background, No 84025 is making its way to Dover from Folkestone. The route along the foot of the cliffs was opened by the South Eastern Railway in 1844. No 84025 was new to Ramsgate on 16th April 1957, serving there for 18 months before relocating to Brighton. A change of regions followed in September 1961 when it transferred to the London Midland's Newton Heath depot. A subsequent transfer saw it at Bolton, which was to be followed by a move back to the Southern Region at Eastleigh – one of those destined for the Isle of Wight. This was not to be and the transfer was cancelled and No 84025 was destined for 'immediate withdrawal' that took place on 11th December 1965.

Bottom right. No 84026 awaits departure time at Henfield during 1961. The station was on the Steyning line, that connected the market town of Horsham with the port of Shoreham-by-Sea, with connections to Brighton. It was built by the London, Brighton & South Coast Railway, and opened from Partridge Green to Shoreham Junction on 1st July 1861, and from Itchingfield Junction to Partridge Green on 16th September 1861. The station equipped with a siding that received coal to serve the Steam Mill and Gas Works. It was used in World War 2 as the loading point for locally grown sugar beet to be transported north to London. Betley Bridge, where the line crossed the River Adur about a mile to the north, was a strategic target for German bombers. The station lost its goods traffic on 7th May 1962 and closed to passenger as a result of the Beeching axe on 7th March 1966 and now forms part of the Downs Link path. No 82026 was delivered from Darlington Works on 18th April 1957, arriving at Ramsgate where it served before moving to Ashford two years later. May 1961 saw it at Brighton where it left that September for Newton Heath, moving two months later to Bolton. It was transferred to Horwich Works, for use as a shunter, in May 1963, returning to capital stock at Bolton that December. A final move to Stockport Edgeley took place in December 1964. It was one of those destined for the Isle of Wight, but as with No 84025 was destined for 'immediate withdrawal' when the scheme was cancelled. Withdrawn on 11th December 1965, it joined No 84025 in being scrapped at Parkgate by Arnott Young's' cutters the following March. *N. Stead collection (207846)*

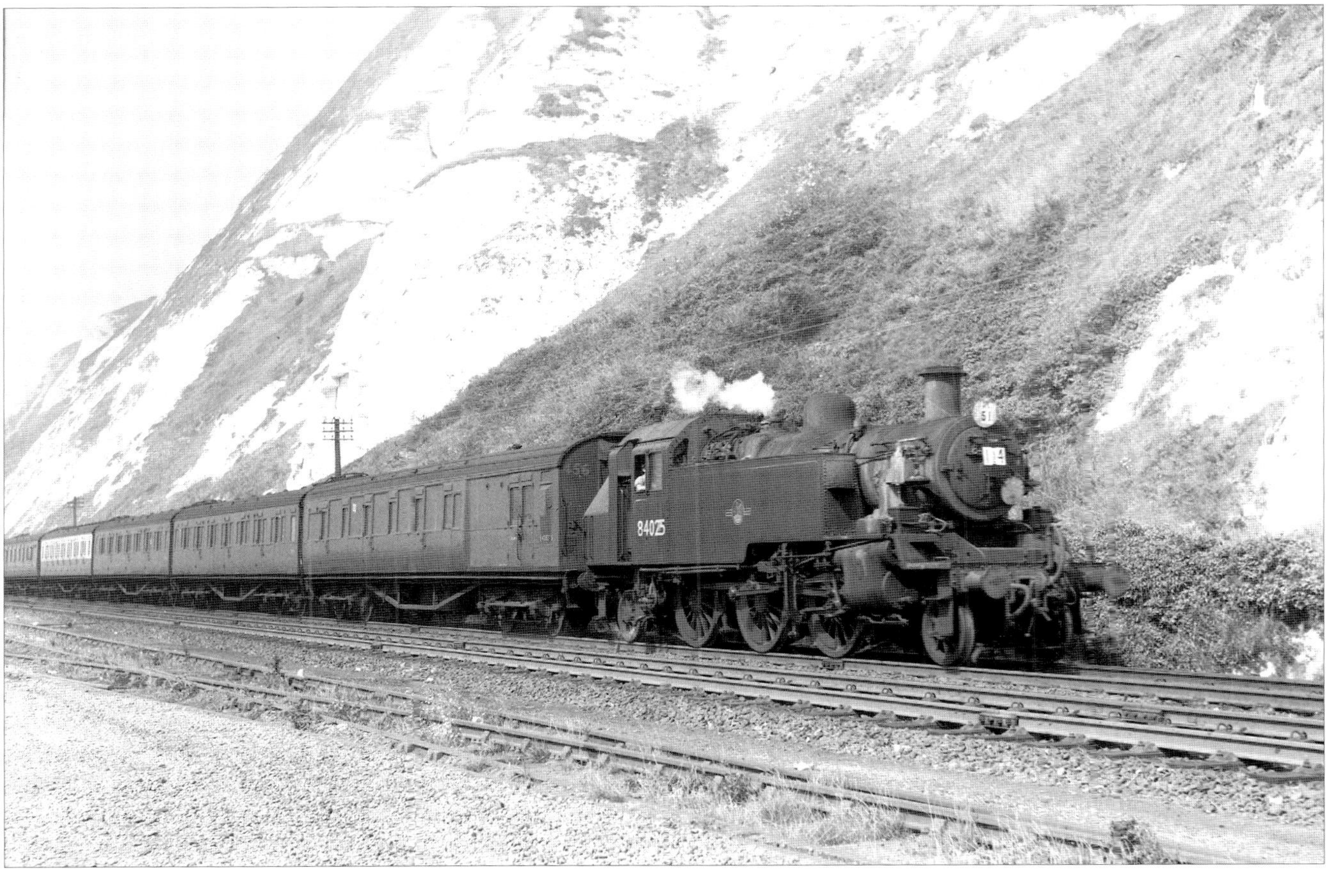
84025

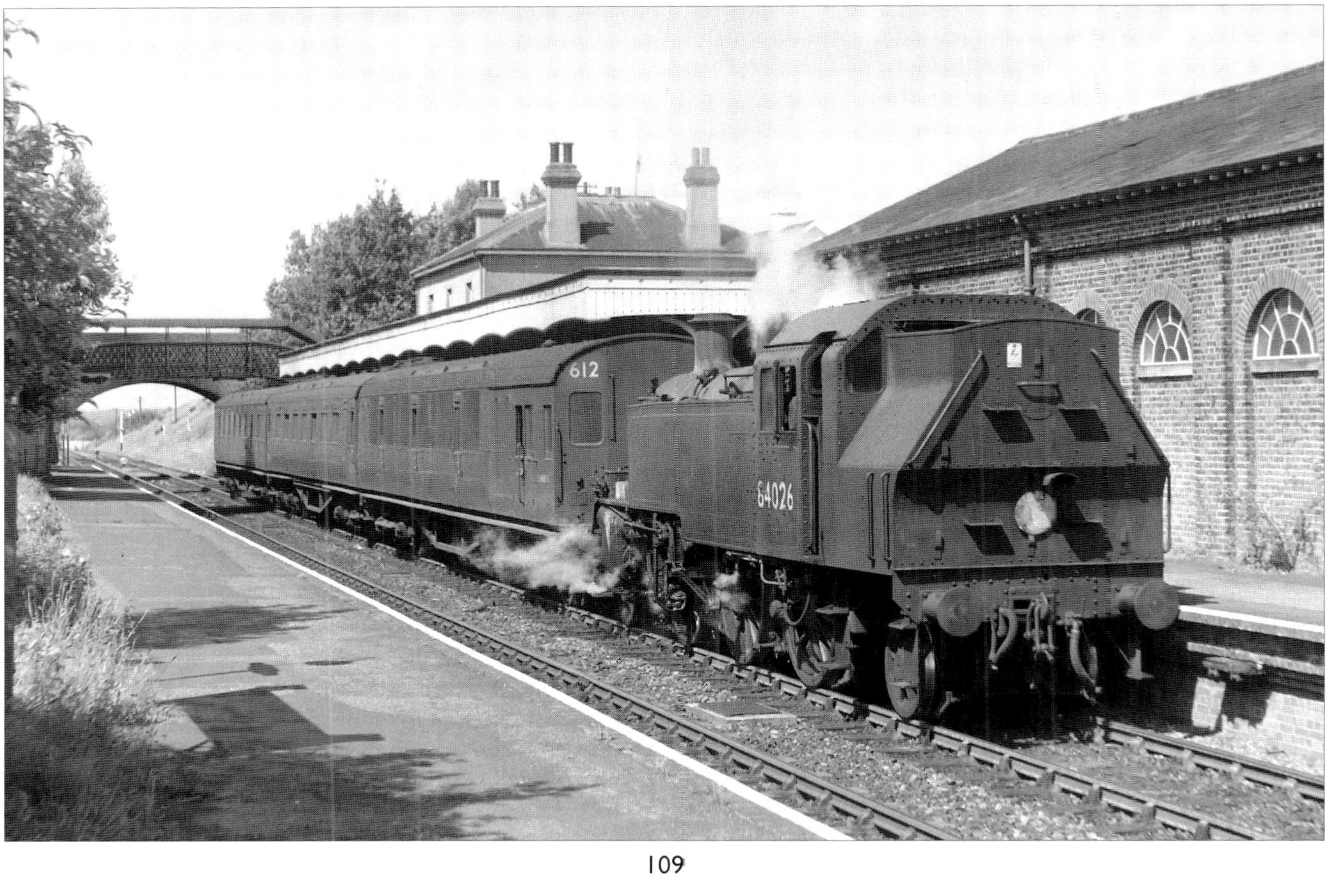
612
84026

Above. The 10.30am from Brighton is seen approaching Horsham on 27th August 1961 behind No 84027. New to traffic on 16th May 1957, it was allocated new to Ramsgate. Two years later it was at Ashford before a move to Brighton in May 1961. A change of regions occurred in September 1961 when it was transferred to the London Midland's Newton Heath depot. Moves to Bolton and Wellingborough followed with a final allocation to Annesley in March 1962. Withdrawn on 2nd May 1964, this was the shortest lived of the Class 2s, serving for 6 years, 11 months and 16 days. *B. Wadey*

Top right. In the pre-multiple unit era, one constant was the marshalling and servicing of coaching stock, with movements to and from the carriage sidings. No 84028 is carrying out such a manoeuvre at Margate on 28th March 1959 whilst allocated to Ramsgate depot. Set No 230 consisted of two Brake Third Corridor coaches sandwiching Composite Corridor, Nos S2763S, S5673S, and S2764S, which was disbanded in December 1959. The locomotive departed Ramsgate in early summer 1959 when it was transferred to Ashford, followed by a move to Eastleigh in May 1961. A move to the London Midland Region followed in September 1961 when it was allocated to Skipton. The failed paper transfer return to Eastleigh for the Isle of Wight saw its withdrawal from Skipton on 11th December 1965, destined for a trip to the Wigan scrapyard of the Central Wagon Co. *R. C. Riley (13087)*

Bottom right. Standing brand new at Darlington Works on 11th June 1957, No 84029 will serve its masters for 7 years and 2 days before consigned to Cashmores at Great Bridge for dismantling. New to Ramsgate, it saw service at Ashford and Eastleigh before transfer to the London Midland Region at Bedford in June 1961. This was followed by periods at Neasden, Kentish Town and Wellingborough before a final move to Leicester Midland from where it was withdrawn on 13th June 1964. Darlington Works was established in 1863 by the Stockton & Darlington Railway, the first new locomotive was built at the works in 1864. Though the railway had amalgamated with the North Eastern Railway (NER) in 1863, it continued to build its own designs for a number of years. In 1877, the first NER designs appeared. Under the LNER it continued to play a major role, producing a new engine each week, and by 1927 the works was the town's largest employer. After nationalisation, the works built both steam and diesel locomotives, including BR Standard Class 2s. In 1954, during the modernisation of BR, the works was enlarged and had grown to cover over 238,000sq ft (22,100m²), but in 1962 the BR Workshops Division was formed and, with rationalisation, the works was run down and closed in 1966.

No 80013 awaits the torch in King's scrapyard in Norwich. New from Brighton Works on 28th August 1951 it entered traffic at Tunbridge Wells West shed, staying there until March 1959 when it moved down to the coast at Brighton. A move westwards saw it relocated to Bournemouth during June 1964 where it remained until withdrawn on 19th September 1966. It was the sole example of the class to travel to Norwich where it was recycled during November the same year. *Ian C. Allen (BR203)*

# Bibliography and Further Reading

Bond, Roland C., *A Lifetime with Locomotives*, Goose & Son Publishers, 0900404302, 1975

Cox, E. S., *British Railways Standard Steam Locomotives*, Ian Allan, no ISBN, 1966.

Gammell, C. J., *Southern Region Engine Workings*, OPC, 9780860935100, 1994

Griffiths, R. & Smith, P.; *The Directory of British Engine Sheds and Principal Locomotive Servicing Points: Vol 1*, OPC, 9780860935421, 1999.

Griffiths, R. & Smith, P.; *The Directory of British Engine Sheds and Principal Locomotive Servicing Points: Vol 2*, OPC, 9780860935483, 2000.

Grindlay, Jim, *British Railways Steam Locomotive Allocations 1948-1968: Part 5 BR Standard & Ex War Department 70000-92250*, Modelmaster Publications, 9780954426253, 2006.

Haresnape, B., *Ivatt & Riddles Locomotives: A Pictorial History*, Ian Allan, 0711007950, 1977

Harris, M. L., (Ed), *Locomotives Illustrated: 21 – BR Standard Tank Locomotives*, Ian Allan.

Longworth, H.; *British Railways Steam Locomotives 1948-1968*, OPC, 9780860935933, 2005.

Longworth, H.; *British Railways Steam Locomotive Allocations*, OPC, 9780860936428, 2011.

RCTS, *A Detailed History of British Railways Standard Steam Locomotives: Vol 1 – Background to Standardisation and the Pacific Classes*, RCTS, 0901115819, 1994

RCTS, *A Detailed History of British Railways Standard*

*Steam Locomotives: Vol 3 – The Tank Engine Classes*, RCTS, 9780993490859, 1997.

RCTS, *A Detailed History of British Railways Standard Steam Locomotives: Vol 5 – The End of an Era*, RCTS, 0901115973, 2012.

Walmsley, T., *Shed by Shed, Part One, London Midland*, St Petroc InfoPublishing, 9780956061553, 2010.

Walmsley, T., *Shed by Shed, Part Two, Eastern*, St Petroc InfoPublishing, 9780956061560, 2010.

Walmsley, T., *Shed by Shed, Part Three, North Eastern*, St Petroc InfoPublishing, 9780956061539, 2010.

Walmsley, T., *Shed by Shed, Part Four, Scottish*, St Petroc InfoPublishing, 9780956061577, 2011.

Walmsley, T., *Shed by Shed, Part Six, Western*, St Petroc InfoPublishing, 9780956061522, 2009.

Williams, A., *BR Standard Steam Album*, Ian Allan, 0711010102, 1980

–, *abc British Railways Locomotives: Combined Volume, Winter 1955/56*, Ian Allan 9780711005068, reprinted 1999

–, *abc British Railways Locomotives: Combined Volume, Summer 1957*, Ian Allan 9780711038455, reprinted 2016

–, *abc British Railways Locomotives: Combined Volume, Summer 1958*, Ian Allan 9780711037694, reprinted 2013

Amongst many others the following websites were useful during preparation of this title:
*www.brdatabase.info*
*www.gracesguide.co.uk*